R. S. ENGELBRECHT

Spring 1962

D0934958

Microbiology for Sanitary Engineers

McGRAW-HILL SERIES IN SANITARY ENGINEERING AND SCIENCE

ROLF ELIASSEN, *Consulting Editor*

McKINNEY · Microbiology for Sanitary Engineers
SAWYER · Chemistry for Sanitary Engineers
STEEL · Water Supply and Sewerage

Microbiology for Sanitary Engineers

Ross E. McKinney

Department of Civil Engineering
University of Kansas

McGRAW-HILL BOOK COMPANY, INC.   1962

New York   San Francisco   Toronto   London

MICROBIOLOGY FOR SANITARY ENGINEERS

45180

Dedicated to Dick, Jim, and Perry
and all my students in
Sanitary Microbiology and Biochemistry

Preface

One of the most unusual aspects of sanitary engineering in the United States is the fact that the engineers responsible for making this country the most sanitary in the world do not have a real understanding of the microbiology of the very processes they design. Waste treatment plants which base their entire operations on the microorganisms within them have been designed for the past fifty years with almost no consideration for the biochemical reactions brought about by the various microorganisms. The net result of ignoring the microorganisms has been retarded development of new biological waste treatment processes.

Some sanitary engineers feel that microbiology is for microbiologists and that the microbiologists should have told the engineers of the use of microorganisms. Unfortunately, the microbiologists who have wandered into this field do not understand the engineering aspects of treatment plant design and operation. Thus, the microbiologist has been unable to tell the sanitary engineer what is needed and the sanitary engineer has been unable to translate the microbiologists' work into practical design.

It is obvious that there is a need either to teach the microbiologist engineering or to teach the engineer microbiology. Unfortunately, sanitary microbiology does not always follow classical microbiology. Classical microbiology is that of pure cultures in concentrated nutrient substrates, while sanitary microbiology is that of mixed cultures in very dilute nutrient substrates. The classical microbiologist often finds himself as ignorant as the engineer in the area of dilute microbiology, and so efforts must be directed to teaching the engineer something about microbiology as it affects the processes that he is most concerned with.

This book was designed to teach future sanitary engineers at M.I.T. It was written by an engineer who became a microbiologist in order to design better waste treatment systems. It is not a polished book that gives the student all the answers but is a rough book that leaves loose ends here and there to be filled in at a later date when more knowledge becomes available. This is a growing book which contains ideas and concepts that will probably be outmoded in a few years but it also contains many that will endure. It is not a compilation of research data from fundamental microbiology or from sanitary engineering. It is an opinion based on the experience of the author, and hence is subject to the prejudices and shortcomings of the author—and he has many. Its value, if it has any, lies in the fact that it is an interpretation of many ideas by many people as examined in the light of the experience of one person.

It is the purpose of this book to examine the fundamental microbiology and the biochemistry of the microorganisms of importance to the sanitary engineer and to interpret their relationship to the various designs which interest the sanitary engineer. It is hoped that this book will assist the sanitary engineer to have a better understanding of what he is trying to do, why his designs work the way they do, and why they do not work as he had hoped. If it stimulates interest in a greater consideration of microorganisms as a tool for the sanitary engineer, it will have accomplished its purpose.

No book is the product of one person and this book is no exception. The major credit for this book goes to Professors I. W. Santry, M. P. Horwood, C. N. Sawyer, and R. Eliassen, whose stimulation and encouragement aroused my desire to understand how microbiology affected sanitary engineering design. From my students came the stimulation required to put down into printed form a series of ever-changing notes for a course which was never taught the same way twice. My secretary, Dorothy Dillon, should receive credit for preparing the manuscript, checking the English, and correcting my spelling. Lastly, it is only right that appreciation be extended to my family for allowing me to take so much of their time to write this book.

Ross E. McKinney

Contents

Microbiology for Sanitary Engineers

PART I

Fundamental Microbiology

CHAPTER 1

Sanitary Microbiology

Sanitary microbiology is a relatively new field which has developed from bacteriology, mycology, phytology, virology, protozoology, and zoology. It is an applied science related to the use of microorganisms in the field of sanitary engineering. Recent advances in biological waste treatment systems for industrial wastes, as well as for domestic sewage, have served as the impetus for establishing sanitary microbiology as a distinct field of its own rather than remaining as an adjunct to medical microbiology.

What Is Sanitary Microbiology?

Sanitary microbiology is a highly specialized phase of microbiology dealing with microorganisms commonly found in water and soil. Sanitary microbiology is usually divided into two general regions, one dealing with water supply and purification and the other dealing with waste disposal. It is readily recognized that the two regions of sanitary microbiology have considerable overlapping and that such a division is an arbitrary one based on the background of the microbiologists working within those regions.

The water microbiologists are primarily concerned with the production of a safe, palatable water for domestic consumption. Their job is to examine the raw water for pathogenic microorganisms and to determine what type of treatment is required to ensure that the water quality is satisfactory from its point of origin to the point of consumption. Originally, the water microbiologists were concerned only with the enteric bacteria, but today they must realize that viruses also can be transmitted through water. The water microbiologists determine the efficiency of each unit process used in water purification with respect to the removal of the various types of pathogenic microorganisms.

Not all of their work is concerned with pathogenic bacteria. Some of it is concerned with nuisance microorganisms which cause tastes and odors and pipe clogging and corrosion. Algae and actinomycetes have been implicated as the causative microorganisms in tastes and odors imparted to water from surface supplies. The water microbiologist must understand why these microorganisms grow and how to control excessive growths. The old adage of "an ounce of prevention is worth a pound of cure" ranks as the most important criterion of operation for the water microbiologist. He must be able to foresee trouble from the nuisance microorganisms and to control them at the source rather than putting the burden on the sanitary engineer at the purification plant. Iron oxidizing bacteria and certain slime producing bacteria have caused problems in sand filters and in distribution pipes. In some instances the biological growths have completely sealed distribution pipes and prevented the flow of water. It has been shown that external corrosion of pipes is accelerated by microorganisms and that even the internal corrosion of water mains has microbiological significance.

The waste disposal microbiologist is concerned not only with water microbiology but also with soil microbiology. Pathogenic microorganisms hold only a small interest for the waste disposal microbiologist, as his major interest lies in the degradation of organic compounds found in both liquid and solid wastes. He has found that microorganisms can completely stabilize the organic matter in liquid wastes, yielding a water which can be reused in industrial processes or even for domestic consumption if properly treated. Solid waste can be stabilized with the production of humus materials which can be used to recover waste land or to enrich tillable soils. The waste disposal microbiologist must understand how microorganisms grow in mixed biological populations and what environmental factors affect the microbial reactions. Unlike his counterpart, the water microbiologist, the waste disposal microbiologist determines the best design criteria to produce a given set of results and gives the information to the sanitary engineer to convert into final designs. Since the liquid effluent from waste disposal plants must find its

FIG. 1-1. Sudbury reservoir in Sudbury, Mass., a typical surface water supply.

FIG. 1-2. Trickling-filter sewage treatment plant at Marlboro, Mass.

way into streams and lakes, the waste disposal microbiologist is also an expert on stream microbiology.

Who Are Sanitary Microbiologists?

There are relatively few sanitary microbiologists who are well trained in all phases of this field. The water microbiologists consist primarily of those microbiologists who received formal training in bacteriology. Their job is the cultivation and identification of pathogenic bacteria or indicator organisms, requiring the technical skills of a bacteriologist. Work with the nuisance organisms usually requires a branching out from the formal bacteriological techniques, but the isolation and identification procedures are closely related so that little additional effort is required to solve problems connected with nuisance organisms.

On the other hand, the waste disposal microbiologist has not evolved from bacteriology but rather from sanitary engineering. Much of the work of the waste disposal microbiologist is concerned with the biochemistry of mixed microbial populations in dilute organic systems. The bacteriologist who has received formal training primarily with pure cultures in concentrated organic solutions finds waste disposal microbiology all but an entirely new science, with engineering and biochemistry of greater importance than conventional bacteriology.

Thus it is that we find the sanitary microbiologists have received their formal education either in bacteriology or in sanitary engineering with the bacteriologist becoming more proficient in engineering as his experience develops and the sanitary engineer taking the rudimentary microbiology obtained in school and developing it further with each sanitary engineering project. There is no such thing as a formal education for a sanitary microbiologist unless experience is considered a part of the formal education system. Like the field of sanitary microbiology itself, the sanitary microbiologist is a hybrid, no two being exactly alike.

Microorganisms of Importance

The sanitary microbiologist is concerned with all groups of microorganisms and even with some which would best be classified as macroscopic, i.e., seen with the naked eye, rather than microscopic. The microorgan-

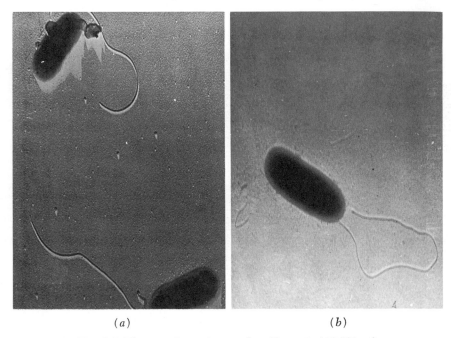

(a)   (b)

FIG. 1-3. Electron photomicrographs of bacteria (20,000 ×).

isms are generally classified into two major groups, *plants* and *animals*. The plants are made up of virus, rickettsiae, bacteria, fungi, and algae; while the animals consist of protozoa, rotifers, and crustaceans. It is the latter two groups of animals that sometimes are classified as macroscopic rather than microscopic.

Plants. The bacteria are the basic units of plant life, being single-cell microorganisms which take in soluble food and convert it to new cells.

Most of the bacteria utilize organic matter for their food, although a few specialized bacteria can utilize inorganic compounds instead of organic compounds. A few bacteria have photosynthetic pigments but these bacteria are of little importance in sanitary microbiology. The bacteria exist as individual cells, in chains, and in clusters. It was originally thought that the bacteria were the smallest plant cells, but both the rickettsiae and the viruses are smaller.

The rickettsiae are plant cells which resemble bacteria in appearance but are much smaller. The rickettsiae are intracellular parasites of fleas, lice, ticks, and mites and many have been shown to be pathogenic to man. Their interest to sanitary microbiologists lies in the fact that the sanitary engineer is often charged with rodent control programs which prevent rickettsial diseases by control of the carrier.

The viruses are the smallest plant cells known at the present time. Like the rickettsiae the viruses are intracellular parasites, deriving their nutrients from the host organisms, and are very close to being considered as pure organic compounds with the power of reproduction within a host organism. Viruses are highly specific in their reactions, with some parasitic to plants and others parasitic to animals. The viruses which are parasitic to bacteria have been designated as bacteriophages. The viruses are too small to be seen by the optical microscope and must be observed through the electron microscope.

On the other side of bacteria in size, the fungi represent the microscopic plants which are lacking in chlorophyll (a photosynthetic pigment which permits the conversion of sunlight energy into chemical energy) and which do not reproduce by binary fission. Actually, bacteria are fungi in the strictest sense but for purpose of simplicity of study the fungi usually exclude the bacteria. Most of the fungi are multicellular microorganisms with branched structures and which pass through several phases in the course of their

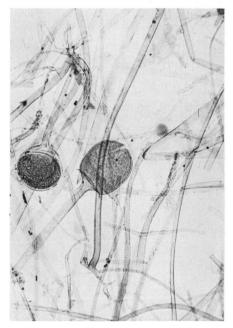

Fig. 1-4. Photomicrograph of the fungi, *Aspergillus* (4,000 ×).

life cycle. The fungi utilize organic compounds as their source of food; in fact, fungi can be found to metabo-

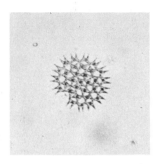

FIG. 1-5. Photomicrograph of the algae, *Pediastrum* (400 ×).

lize almost every type of organic compound which occurs in nature. It has been estimated that there are over 100,000 distinct species of fungi which have been studied to date. The large number of fungi has had a retarding effect on most microbiologists' interest; but their importance in sanitary microbiology requires only an understanding of the biochemistry of the fungi without a need for the entire taxonomy.

The algae consist of all the microscopic plants which have chlorophyll but lack definite stems and leaves as in the case of higher plants. Some of the algae are monocellular, while others are multicellular. Like the fungi, the algae have been studied more from a taxonomy viewpoint than from a biochemical view. Much of the interrelated biochemistry of the algae and the other microorganisms has still to be studied in sanitary microbiology.

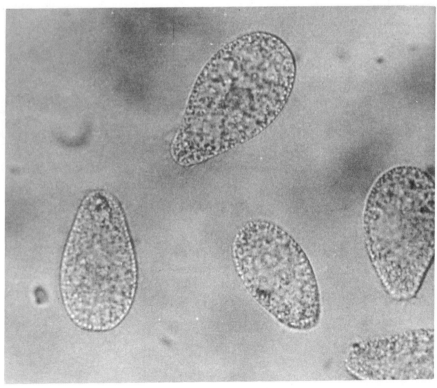

FIG. 1-6. Photomicrograph of the ciliated protozoa, *Colpidium* (4,000 ×).

Animals. The basic animal units are the protozoa which are single-cell animals. Most of the protozoa have the ability to metabolize solid food by means of complex digestive systems but there are a few protozoa which appear to be related to plants as well as to animals. Most protozoologists consider the single-cell, flagellated microorganisms containing chlorophyll as protozoa rather than algae. The microbiologist who must view all groups of microorganisms considers the chlorophyll-containing microorganisms as algae, while the single-cell, flagellated microorganisms which must utilize soluble food as a plant, but do not contain chlorophyll, are considered as protozoa. As always in the classification of microorganisms there are some transition forms which have the characteristics of two major groups and which the microbiologist must classify arbitrarily. The importance of the protozoa lies in the fact that their major source of food are the microscopic plants.

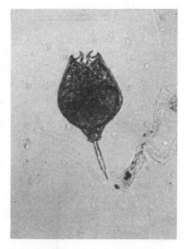

Fig. 1-7. Photomicrograph of rotifer (400 ×).

The rotifers are more complex animals than the protozoa, being multicellular rather than monocellular. Their body is still flexible like the protozoa but one of their characteristics lies in the action of the cilia, short, hairlike appendages. The rotifer's cilia are located on its head and serve as its organ of motility, as well as the means for obtaining food. These cilia are located in two circular rows and move in a manner that gives the appearance of two counterrotating wheels and hence the name rotifer.

Crustaceans are actually macroscopic animals but a few crustaceans can be best observed through the microscope and hence are included with the microscopic organisms. The microscopic crustaceans are multicellular organisms which have a hard shell to protect the vital organs. The crustaceans are important in sanitary microbiology as predators on the microscopic plants and as indicators for toxicity.

Biochemistry

Sanitary microbiology is more than the study of microorganisms; it is also the study of the biochemical reactions which they bring about. The identity of the various microorganisms responsible for most sanitary engineering projects takes a secondary role to the total reaction produced by the entire mass of microorganisms. For this reason the sanitary microbiologist is more of a biochemist than a conventional microbiologist. Fortunately for the sanitary microbiologist he is not interested in the highly specialized biochemical reactions of the unique or the unusual microorganisms, as is often the case in conventional microbiology. He deals with the common everyday, garden variety of microorganisms which can metabolize the organic matter to completion rather than to just a limited extent. The biochemical reactions of the bacteria, the fungi, the algae, the protozoa, the rotifers, and the crustaceans all follow the same general patterns with the same general quantitative relationships. Thus, once the biochemistry of one group of microorganisms has been mastered, it is easy to expand this information to include all groups of microorganisms.

Fundamental and Applied Microbiology

Sanitary microbiology is an applied field which is concerned with the solution of practical problems in sanitary engineering. Because of its practical implications the need for a thorough understanding of fundamental microbiology has been questioned. It is my belief that the best solution to the practical problems in sanitary engineering can be obtained only through a complete understanding of fundamental microbiology and the application of these fundamentals to the particular problem in question. Because of the importance of both the fundamental and the applied aspects of sanitary microbiology, this book is divided into two sections, Fundamental Microbiology and Applied Microbiology. This permits the student to utilize the entire book as a unit or as two separate units as he desires.

CHAPTER 2

Observations of Microorganisms

The chief tool of the microbiologist is the microscope, with which it is possible to see and to study the most minute microorganisms. There are two types of microscopes available to the microbiologist, the optical microscope and the electron microscope. The optical microscope is routinely used for observation of the larger microorganisms in both the living and dead state, while the electron microscope is used to observe the very minute microorganisms in the dead state only. In order to obtain the maximum use from the microscope, it is necessary for the microbiologist to know and to understand the function of every part and to realize its limitations.

Optical Microscope

The optical microscope is shown in Fig. 2-1. This microscope is the common monocular microscope used in most sanitary laboratories. A cross section of the monocular microscope is shown in Fig. 2-2 in which all the major parts are labeled.

An external light source is used to concentrate sufficient light on the specimen to allow the observer to see all the specimen without strain. The microscope lamps range from a simple lamp, as shown in Fig. 2-3, to a large lamp with adjustable diaphragms, as shown in Fig. 2-4. The light from the lamp is reflected into the microscope by a mirror. Usually, the mirror is double-sided, with one side being a flat mirror and the other side being a concave mirror. The flat mirror gives a parallel light source on the specimen, while the concave mirror concentrates the light on the specimen. Both mirrors have their place and the microbiologist should become familiar with the light patterns of each side. The light patterns are shown schematically in Fig. 2-5.

Most microscopes employ condensers to ensure proper illumination in the objective. The condensers usually have diaphragms which permit reduction of the excess light in the objective. Proper use of the condenser and the diaphragm is most important and is the hardest part of the microscope for the student to master. Too much light through the condenser

11

FIG. 2-1. Typical monocular microscope with mechanical stage.

will make some specimens appear invisible, while too little light hides the details of some specimens.

The flat portion of the microscope on which the specimen rests is called the stage. All modern microscopes are equipped with mechanical stages which permit movement of the specimen easily without undue motion. The mechanical stage has two geared movements, forward and backward as well as right and left. The optics of the microscope result in inversion of the image so that the specimen must be moved in the opposite direction to the apparent motion as observed in the microscope. It takes a little practice to realize that to move the specimen to the right in the microscope requires a left movement of the specimen on the stage.

The light passes through the specimen into the objective, the lower lens. Normally, three objectives of different magnification are located on a revolving nosepiece. The standard objectives are 10 ×, 43 ×, and 97 ×. The 10 × and 43 × objectives are used with air between the lens and the

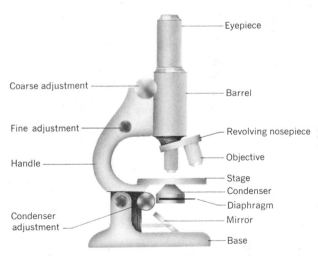

FIG. 2-2. Schematic diagram of optical microscope.

specimen, while the 97 × objective uses oil between the lens and the specimen.

FIG. 2-3. Simple microscope lamp.

From the objective the light travels up through the barrel into the eyepiece where the image is magnified even further. The two common eyepieces have 5 × and 10 × magnification. The total magnification of the image is the product of the eyepiece times the objective. A 10 × eyepiece with a 43 × objective will give 430 × magnification.

The barrel is attached to the body of the microscope through a series of gears. One set of gears is attached to the coarse adjustment, while the other set is attached to the fine adjustment. The coarse adjustment is used to locate the approximate focus of the specimen, while the fine adjustment permits final focusing.

The latest design of microscopes is shown in Fig. 2-6. This microscope has a built-in light source with permanent alignment of the axis of the light source with the axis of the optical system. There are four objectives, a 3.5 × in addition to the three standard objectives previously listed. The barrel of the microscope is built as part of the body. Instead of moving the barrel back and forth for focusing, the stage moves. This prevents damage to the optical align-

FIG. 2-4. Typical compound microscope lamp.

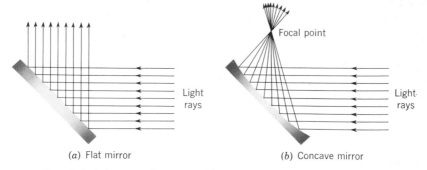

(*a*) Flat mirror   (*b*) Concave mirror

FIG. 2-5. Schematic diagrams of light patterns with optical microscope.

ment. The controls have been placed for more convenient use by the microscopist. The new binocular microscopes are a definite advance over the old style; but the permanence of microscopes will prevent immediate replacement of the old-type microscopes with the new ones.

Focusing the Microscope

One of the more difficult things to do with a microscope is to find the

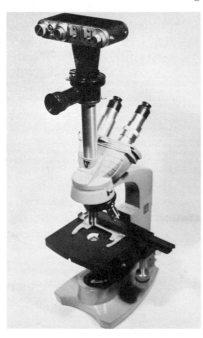

FIG. 2-6. Modern binocular microscope with built-in light source and equipped with 35-mm camera.

microorganisms that you are looking for. The easiest way to locate the microorganisms is to place the specimen on the stage with the optical axis of the microscope passing through the specimen. The $10 \times$ objective is usually used with the $10 \times$ eyepiece to give $100 \times$ magnification. The objective is lowered until it almost touches the specimen. With most microscopes the $10 \times$ objective will stop short of the specimen automatically. Slowly the objective is moved away from the specimen with the coarse adjustment. Once the specimen comes into view, the fine adjustment is used to make the precise focusing. If you wish to examine the specimen at $970 \times$ magnification, a drop of immersion oil is placed on the specimen and the nosepiece revolved until the $97 \times$ objective locks in place. If the microscope is properly made, the specimen will be

almost in perfect focus so that only minor focusing with the fine adjustment will be required. If the shift from the 10 × objective to the 97 ×
objective does not give an approximate focus, it is an indication that the
objectives are not properly seated in the nosepiece. Care must always be
use. If ~~~~ wipe the oil from the 97 × objective with lens paper after each
~~sion oil dries on the lens, it will be impossible to obtain accurate focusin

Limit of Resolut

The ability to magnify ~~~ ~~~~ is not the only factor which determines the sensitivity of the microscope. Clarity is as important as magnification, if not more important. The ability of a microscope to show
small microorganisms clearly and distinctly is determined by its limit of
resolution. The limit of resolution is defined as the distance between two
adjacent points which still give distinct images. If very small objects are
to be seen clearly, the limit of resolution must be small. Three factors
determine the limit of resolution: (1) the wavelength of light, (2) the
refractive index of the medium between the specimen and the objective,
and (3) the aperture of the objective.

The refractive index of the medium between the specimen and the
objective and the aperture of the objective are functions of the construction of the objective and are combined into a single number called the
numerical aperture (N.A.). Most manufacturers stamp the numerical
aperture on each objective to prevent misunderstanding as to the construction of the objective. The limit of resolution can be expressed mathematically as a function of the wavelength of light and the numerical
aperture.

$$\text{Limit of resolution} = 0.6 \text{ (wavelength of light)}/\text{N.A.}$$

There is a definite limit to the value of N.A. The dry objectives using
air as the medium have N.A. values less than 1.0, while the oil immersion
objectives usually have a N.A. value under 1.5. Since the values of N.A.
are fixed for a given objective, the limit of resolution is related to the
wavelength of light. If we consider the smallest wavelength of visible
light and the maximum N.A., we find that the limit of resolution cannot
be less than 0.17 micron for the optical microscope. This means simply
that we cannot use the optical microscope to see objects smaller than 0.17
micron. Thus it is that the optical microscope using visible light has a
limit of magnification which cannot be exceeded. This is the reason why
virus particles cannot be seen in the optical microscope.

Electron Microscope

The microscopist has not been content to stop at the magnifications obtainable with the optical microscope and has turned toward electronics to find a more powerful microscope. The principle of the electron microscope is very similar to that of the optical microscope. A schematic diagram in Fig. 2-7 shows how the electron microscope works.

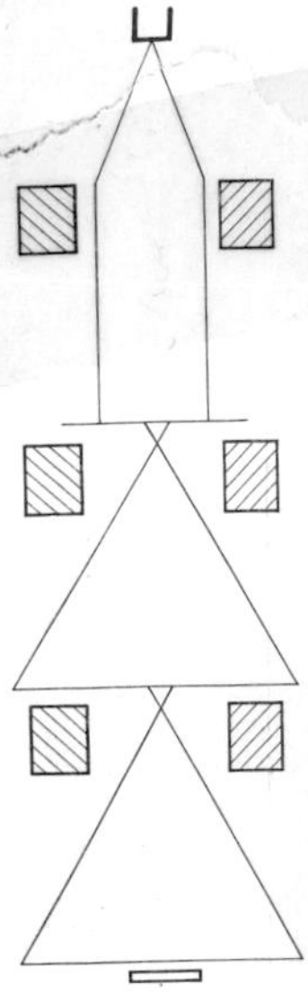

The cathode discharges electrons with an accelerating potential of 50,000 volts. Since these electrons would be easily absorbed by matter in air, the electrons must travel in a vacuum. A magnetic condenser between the cathode and the specimen aligns the electrons so that they pass uniformly through the specimen. Differences in density of specimen causes some of the electrons to be absorbed by the specimen. Once past the specimen the electrons are deflected by a magnetic objective similar to light waves passing through the optical objective. An intermediate image projector magnifies a portion of the electrons and gives a greater magnification. The electrons cannot be seen directly but produce a visible image on a fluorescent screen in the same manner as a television tube.

Electron source

Magnetic condenser

Specimen

Magnetic objective

Projector

Image screen

Fig. 2-7. Schematic diagram of electron microscope.

Since the brightness of the image produced on the fluorescent screen is dependent upon the electrons which hit it, it is essential that most of the electrons pass through the specimen. It is also desired that the major absorption of the electrons be by the specimen so that maximum contrast can be obtained between the specimen and the background. Ideally, the specimen should be maintained unsupported in space but this is actually impossible. It does mean that the specimen must be mounted on a very thin film with low electron-absorbing properties. In the early days the specimens were coated with gold and other metallic surfaces to give strong contrast but newer techniques have permitted slicing of bacteria and examination of the internal structures without metallic coatings.

The limitations of the electron microscope are similar to those of the optical microscope. The major difference between the two systems is that the wavelength of the electron is much shorter than that of light. It is also possible to build electronic lenses with higher numerical apertures than optical lenses. At the present time the limit of resolution of the electron microscope is 0.0003 micron, making it possible to examine the minutest objects. In fact, the electron microscope can magnify some microorganisms so large that it is of no value. The major uses of the electron microscope are in observation of viruses and in the study of cell structure within the bacteria.

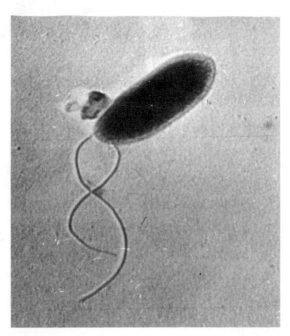

Fig. 2-8. Electron photomicrograph of a bacterium (40,000 ×).

The electron microscope is not a common laboratory tool and much work is still needed before it can be simplified to the point of routine use in sanitary microbiology. At the present time the complexity of the equipment requires a special electronics technician as well as a microscopist. In view of the rapid developments in the electronics field, there is little doubt that the electron microscope will become an important tool in the sanitary microbiology laboratory.

Phase Microscopy

The attainment of maximum magnification in the optical microscope caused attention to be focused on other aspects of microscopy. One of the latest innovations is phase microscopy which increases the contrast in various portions of living cells and permits differentiation of parts within the living cell.

Phase contrast is produced by special equipment inserted into the microscope to cause a portion of the light passing the specimen to be out of phase with the remainder of the light. As the light passes through the specimen, its phases change again in proportion to the density of the

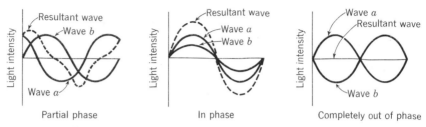

Partial phase   In phase   Completely out of phase

FIG. 2-9. Schematic representation of light patterns in phase-contrast microscope.

material through which the light is passing. In some areas the two phases come back together and produce a brighter light, while in other areas they become further out of phase and hence are darker. Phase contrast is used to note small differences in cellular structures but has had limited use in sanitary microbiology.

Dark Field

Dark-field illumination is an old technique used in microscopy that has

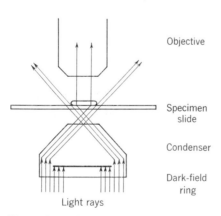

FIG. 2-10. Schematic representation of dark-field microscopy.

some value in sanitary microbiology. A dark-field ring is placed in the condenser between the light source and the specimen and prevents the light from passing directly into the objective. As the light passes through the specimen, it is deflected sufficiently so that it passes into the objective. In this way the background appears black while the specimen appears bright. Since most of the light is blocked by the dark-field ring, a bright light source is required for dark-field illumination. Dark-field illumination has its major value in studying bacterial motility where the contrast between living cells and the surrounding medium is very low.

Living Microorganisms

It is very important for the sanitary microbiologist to study microorganisms in their natural environment. The hanging-drop technique was developed to observe bacteria and fungi, while the Sedgwick-Rafter cell was developed to observe the larger microorganisms. A depressed slide is used for the hanging-drop observations. A drop of the living culture is placed on a cover glass which is carefully inverted over the depression

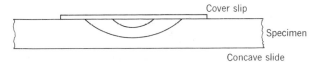

Cover slip

Specimen

Concave slide

FIG. 2-11. Schematic cross section of hanging drop.

in the slide. A thin ring of grease around the depression in the slide will prevent the cover slip from sliding and will eliminate air currents. Care must be taken in focusing, especially at 970 × magnification. It is very easy to depress the objective too far and to shatter the cover slip. Focus should always be away from the cover slip and never toward the cover slip.

Large microorganisms are normally observed at 100 × magnification. Since the density of large organisms is low, it is necessary to use a large sample in order to find the desired microorganisms. The Sedgwick-Rafter cell shown in Fig. 2-13 was developed to hold exactly one milliliter of sample. Its dimensions were selected to give a depth of 1 mm with 20 mm width and 50 mm length. A cover slide is usually placed over the Sedgwick-Rafter cell to prevent movement due to surface agitation. It is possible to use the Sedgwick-Rafter cell for quantitative enumeration of large microorganisms.

Often the concentration of microorganisms in 1 milliliter (ml) of sample is not sufficient to give a satisfactory count. The microorganisms can easily be concentrated by centrifuging the sample and pouring off the clear liquid. If the concentration of microorganisms is too high, quantita-

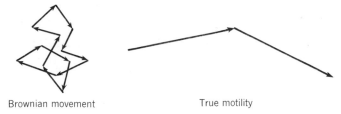

Brownian movement

True motility

FIG. 2-12. Sketch of Brownian movement and true motility.

tive counts can be made by counting several fields rather than the entire sample. Care must be taken so that a sufficient number of fields are counted to yield a statistically valid count.

One of the prime reasons for observing living microorganisms is to determine their motility. It is easy to distinguish motility in the large microorganisms, but bacteria pose a problem. Bacteria are only slightly larger than colloids and exhibit some of the properties of colloids. One of the colloidal properties exhibited by bacteria is Brownian movement. Brownian movement is the random motion of colloids due to molecular collisions. Care must be taken to distinguish between Brownian movement and true motility. Normally, true motility results in a direct path at high velocities, while Brownian movement is a slow random motion with no direct path.

Stains

The most common method for observing bacteria is by the use of stained preparations. The retention of coal-tar dyes by bacteria permits easy visibility of the microorganisms under the microscope. The color of the dye is imparted by its chemical structure. The coal-tar dyes use aromatic ring bases such as benzene, naphthalene, or anthracene. Various chemical groups are attached to the aromatic rings to impart the different colors. The color-producing groups are called chromophores. Certain other chemical groups called auxochromes assist in strengthening the chromophores. The common chromophores include the p-quinone group and the diazo group whose structures are shown in Fig. 2-13. The auxochromes include the amino group ($-NH_2$) and the hydroxyl group ($-OH$).

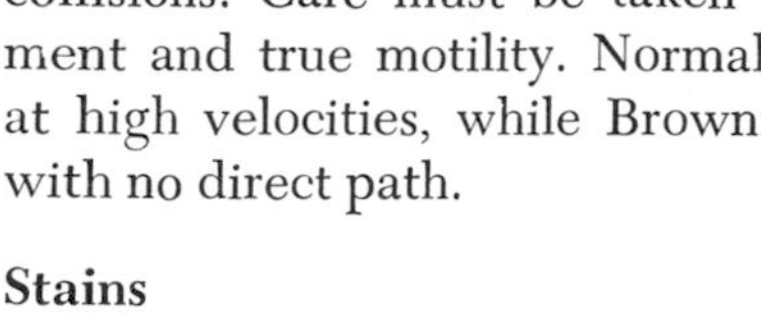

$-N=N-$

$-N=N-$

p-Quinone   Diazo

FIG. 2-13. Chemical structure of the chromophore groups.

Most dyes are organic bases or acids. The acidic or basic properties of dyes permit easy classification. The acid dyes ionize in aqueous solutions to yield a negatively charged dye nucleus, while the basic dyes ionize to form positively charged dye molecules. Neutral dyes are made up of mixtures of basic dyes and acid dyes so that the net electric charge of the mixture is zero, neutral. The final group of dyes has been designated as the indifferent dyes. The indifferent dyes could be better classified as nonionic, as they have no ionizable chemical groups. The structures of six of the commonest dyes are given in Fig. 2-14.

Theory of Staining. Staining of microorganisms by dyes is primarily a physicochemical reaction between the dye molecule and certain chemical groups within the microbial cell. Microorganisms are composed primarily of proteins, lipids, and carbohydrates. Proteins are made up of units of amino acids which ionize in basic solutions to form negative groups and in acidic solutions to form positive groups. The ability of a molecule to ionize to both acidic and basic groups is called amphoteric and is illustrated in Fig. 2-15. Pure lipids do not ionize but rather undergo partial hydrolyses to form free fatty acids which ionize in basic solutions to form negatively charged groups. Carbohydrates do not ionize as they are nonionic polyalcohols.

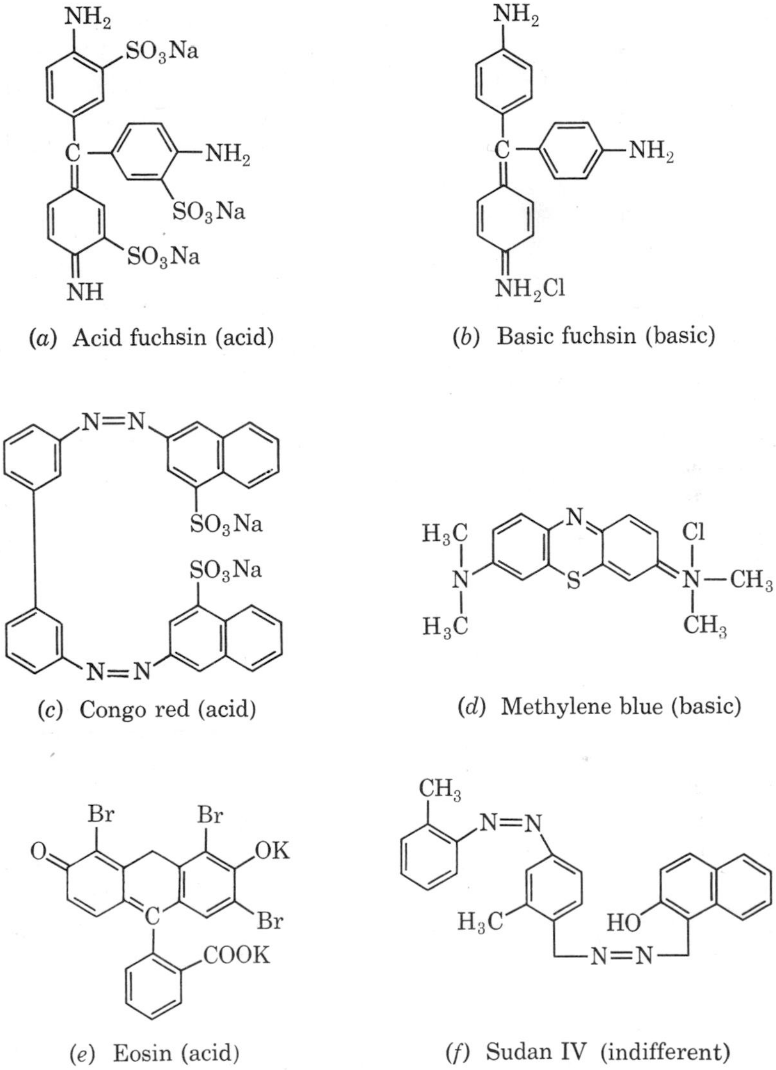

(a) Acid fuchsin (acid)

(b) Basic fuchsin (basic)

(c) Congo red (acid)

(d) Methylene blue (basic)

(e) Eosin (acid)

(f) Sudan IV (indifferent)

FIG. 2-14. Chemical structure of six common dyes.

It is a well-known electrical phenomenon that particles of opposite charge attract, while similarly charged particles repel each other. In basic solutions microorganisms carry a negative electric charge due to the ionization of the carboxyl groups on the amino acids and fatty acids. It is not surprising to find that in basic solutions the microorganisms take up basic dyes so readily and that in acidic solutions the acid dyes are taken up. If the pH of the microbial solution is slowly changed from basic to acidic, the quantity of basic dye taken up by the microorganisms decreases. It has been shown that there is a quantitative uptake of dye in proportion to the degree of ionization.

The indifferent dyes cannot react by electrical attraction because they have no ionizable groups. For the most part the indifferent dyes depend upon their attraction to lipids. The large hydrocarbon molecules forming the indifferent dyes are strongly hydrophobic and dissolve into the lipid portions of the cell. For this reason the indifferent dyes have been used to demonstrate lipid areas in the microbial cells. Some of the ionic dyes are also capable of dissolving into the lipid regions to a slight extent. This permits the microorganisms to retain the color of a basic dye even when they are immersed into an acidic solution. Needless to say, the color retention is slight.

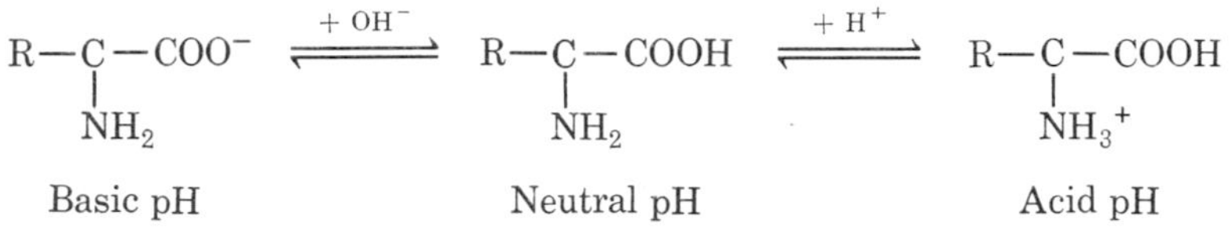

$$R-\underset{\underset{NH_2}{|}}{C}-COO^- \overset{+\ OH^-}{\rightleftharpoons} R-\underset{\underset{NH_2}{|}}{C}-COOH \overset{+\ H^+}{\rightleftharpoons} R-\underset{\underset{NH_3^+}{|}}{C}-COOH$$

Basic pH   Neutral pH   Acid pH

FIG. 2-15. Amphoteric amino acids.

Gram Stain. The Gram stain is an important differential stain for sanitary engineers. The coliform test on which much of the sanitary requirements of water and sewage are based depends upon final confirmation of coliform bacteria with the Gram stain. Essentially, the Gram stain divides bacteria into two groups as a result of their color reactions.

The most common Gram stain used today is Hucker's modification. The reagents are as follows:

1. Crystal violet ammonium oxalate
   Solution A:
       Crystal violet . . . .   2 gm
       Ethyl alcohol (95%) . . . . 20 ml
   Solution B:
       Ammonium oxalate . . . . 0.8 gm
       Distilled water . . . . 80 ml
   After preparing solutions A and B, mix the two solutions to yield a single solution.
2. Iodine solution
       Iodine . . . . 1.0 gm
       Potassium iodide . . . . 2.0 gm
       Distilled water . . . . 300 ml
3. Ethyl alcohol (95%)

4. Safranin
   Solution 1:
      Safranin 0 . . . .   2.5 gm
      Ethyl alcohol (95%) . . . .   100 ml
   Solution 2:
      Solution 1 . . . .   10 ml
      Distilled water . . . .   100 ml

The procedure for the Gram stain is given as follows:

1. Dry smear of microorganisms on clean glass slide.
2. Flame the smear to fix microorganisms to slide.
3. Stain with crystal violet for 1 min.
4. Wash the excess dye from the slide with water and shake off excess water.
5. Flood with iodine solution for 1 min.
6. Wash excess iodine off with water, and blot dry.
7. Decolorize in 95 per cent ethyl alcohol, and blot dry.
8. Counterstain with safranin for 1 min.
9. Wash the excess dye with water, dry, and examine.

The Gram-positive cells retain the purple color, while the Gram-negative cells stain red.

The Gram stain is not an absolute test and should be used with care. It should be used only with 24-hr cultures of bacteria since some bacteria begin to lose their Gram positiveness after this period of incubation. It has been noted that some bacteria are Gram variable, i.e., some of the bacteria will stain red and some will stain purple.

There has been considerable controversy as to why bacteria react as they do to the Gram stain. It has been shown that there is no difference in chemical structure of the Gram-positive and the Gram-negative bacteria, but there appears to be a difference in chemical configuration. The Gram-positive cells appear to have a lower isoelectric point because of the configuration of the nucleic acids. The lower isoelectric point gives a stronger electric charge and a stronger attraction for both basic dyes and cations. The most noticeable attraction for cations is the increased concentration of magnesium. It is felt that the magnesium-iodine-crystal violet tricomplex resists decoloring with alcohol and permits the Gram-positive bacteria to remain the violet color, while the Gram-negative bacteria are completely decolorized. The majority of bacteria of interest to sanitary engineers are Gram-negative so that it is only with the coliform test that the Gram stain is of much value.

Capsule Stain. The most difficult bacteria structure to stain has been the capsule. The bacterial capsule has shown no affinity for the normal coal-tar dyes. Examination of data on the chemical composition of the bacterial capsule has shown that it is primarily a polysaccharide. As already indicated, carbohydrates cannot be stained by the acidic or basic dyes. Two methods have been used to demonstrate capsules: (1) oxidation of the hydroxyl groups with periodic acid to form carboxyl groups, followed by normal basic dyes; and (2) precipitation of metallic flocs such as copper hydroxide and iron hydroxide onto the surface of the capsule, followed by crystal violet with acetic acid to permit attraction of the dye at the metallic interface.

The newest method used for staining capsules is by direct dyeing with Alcian Blue 8GN150, a cotton-textile dye. Although it has not been accepted by bacteriologists as a capsular stain it has had extensive use in sanitary microbiology as such. In a basic solution Alcian Blue reacts with the internal hydroxyl groups of the carbohydrate molecules to form a blue complex. In acidic solutions Alcian Blue reacts as a conventional acid dye. This dual reaction makes it imperative that basic solutions be used when staining for capsules. The reagents required for the Alcian Blue stain are as follows:

1. Alcian Blue solution
   Alcian Blue 8GN150 . . . .   100 gm
   Ethyl alcohol (95%) . . . .   100 ml
2. Dilute Alcian Blue
   Solution 1 . . . .   10 ml
   Distilled water . . . .   90 ml
3. Carbolfuchsin stock
   Solution A:
      Basic fuchsin . . . .   0.3 gm
      Ethyl alcohol (95%) . . . .   10 ml
   Solution B:
      Phenol . . . .   5 gm
      Distilled water . . . .   95 ml
   Mix solutions A and B
4. Dilute carbolfuchsin
   Carbolfuchsin stock . . . .   10 ml
   Distilled water . . . .   100 ml

The procedure for the Alcian Blue stain is as follows:

1. Dry smear of microorganism on clean slide.
2. Flame smear to fix microorganisms to slide.
3. Stain with dilute Alcian Blue for 1 min.
4. Wash excess dye from slide with water.
5. Counterstain with carbolfuchsin for 10 sec.
6. Wash immediately with water, dry, and examine.

The polysaccharide material stains blue while the protoplasm stains red. Under certain conditions, the Alcian Blue will be overstained with carbolfuchsin, giving a purplish color rather than a clear red. Thus, care must be taken in interpreting the results obtained with Alcian Blue. In young cultures of actively motile bacteria, the capsule will be so thin that it will not be seen. As the culture ages, the capsular material will build up and will stain with Alcian Blue. The size of the dye molecule is quite large so that it will not penetrate into living cells. The net result is that Alcian Blue will only stain the external polysaccharide of living cells. Once a cell dies, the porosity of the cell increases and internal polysaccharide can be stained.

Spore Stain. The ability to demonstrate spores is very important in the identification of bacteria. Spores are not easily demonstrated because of

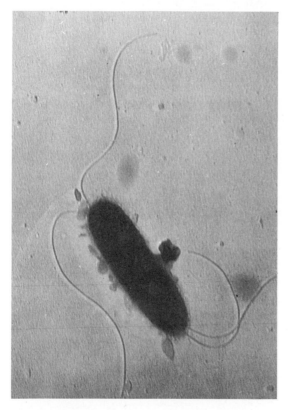

Fig. 2-16. Electron photomicrograph of a bacterium with flagella at both ends of the cell.

their low chemical reactivity. Once again, the problem is related to carbohydrates. The spore coating has been shown to be primarily polysaccharide. It is possible to stain spores directly with Alcian Blue, but the color is not intense. With old cultures spores are easily seen when the microorganisms are stained with the normal coal-tar dyes. The cells take the stain but the spores remain unstained.

Flagella Stain. Demonstration of the type of flagella on bacteria is used as one of the important keys in identification. Yet flagella staining is the most difficult technique there is. The size of flagella is such that they approach the limit of resolution of the optical microscope. In order to see the flagella, a chemical precipitate must be built up on the flagella prior to staining. Alum, tannic acid, and mercuric chloride have been used as fixers to increase the flagella size, with basic fuchsin as the stain. The fragility of flagella as well as the difficulty in the technique has caused most microbiologists to ignore the flagella or to use the electron microscope to show the flagella.

Vital Staining. The microbiologst has long sought a method to demonstrate which bacteria are alive and which bacteria are dead in a given solution. Neutral red in dilute solution has been used for vital staining, with the dead cells being permeable to the dye and the living cells being impermeable.

Methylene blue has been used to stain the nucleus as well as normal protoplasm. When a bacterium ages, the quantity of nucleic acids decreases. The net effect is that the uptake of methylene blue is less in old and dead cells than in young active cells. The use of methylene blue has been of value in demonstrating the active bacteria in activated sludge masses and in demonstrating the aging of higher animals.

SUGGESTED REFERENCE

1. Conn, H. J., "Biological Stains," 6th ed., Biotech Publications, Geneva, N.Y., 1953.

CHAPTER 3

Bacteria

The basic group of microorganisms of importance to the sanitary engineer are the bacteria. These microorganisms are the cause of most sanitary problems. Uncontrolled, the bacteria produce odors and objectionable conditions. Some bacteria attack higher plants and animals and destroy them. It is the sanitary engineer's job to see that the pathogenic bacteria are destroyed before they do damage. He also uses the bacteria under controlled conditions in waste treatment plants to stabilize organic matter and thereby prevent objectionable conditions. The importance of the bacteria is so great that most of this book will deal with bacteria.

Bacteria are the basic plant unit, being the simplest form of plant life. Many people find it difficult to believe that bacteria are plants and not animals. This erroneous impression is derived from the fact that plant life is associated with photosynthesis and the presence of the green pigment, chlorophyll. Most bacteria, especially the ones commonly found in nature, do not contain chlorophyll and are colorless. Bacteria are classified as plants because of their structure and method of food intake. They are single-cell organisms which utilize soluble food. Each cell is an independent organism capable of carrying out all the necessary functions of life.

Distribution

Bacteria are found everywhere in nature; in water, in soil, and in air. Most of the bacteria are found in water and in soil which has a high moisture content, for the bacteria must have an aqueous environment to obtain food. The bacteria in the air are associated largely with particulate materials. Bacteria are distributed in nature according to the presence of nutrient materials used as food. Soil contains all types of bacteria necessary for the degradation of organic matter. If bacteria are needed to stabilize unusual organic compounds, the engineer has only to look under his own feet to find them.

Cytology

In order to know why bacteria can stabilize organic compounds in wastes, how they do it, and how to utilize this information in waste treat-

27

ment design and operations, it is necessary to know and to understand the function of each part of the bacterial cell. Bacteria come in three shapes: (a) rods, (b) spheres, and (c) spirals. The three bacterial shapes are shown in Fig. 3-1. The technical names for these shapes are *bacillus* for rod, *coccus* for sphere, and *spirillum* for spiral.

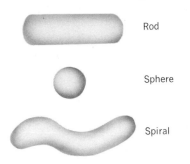

Rod

Sphere

Spiral

Fig. 3-1. Three common shapes of bacteria.

Forms

The forms that these three shapes of bacteria take vary, as shown in Fig. 3-2. The rod is the most common bacterial form and can be observed in three distinct groups: (a) individual cells, (b) diplo or twin cells, and (c) chains of cells. The individual and diplo forms will be most normally observed. A growing bacteria culture will show both forms, since all rods will appear as diplos just before dividing. Chains occur to a limited extent with definite species of bacteria but can be induced by changing the chemical environment.

The spheres have the most different groups: (a) individual, (b) diplo,

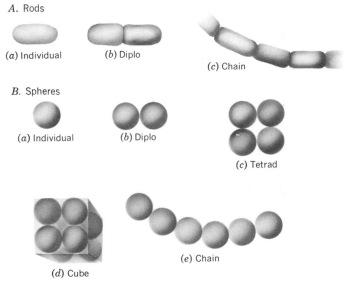

A. Rods

(a) Individual (b) Diplo

(c) Chain

B. Spheres

(a) Individual (b) Diplo

(c) Tetrad

(d) Cube

(e) Chain

Fig. 3-2. Sketch of bacteria forms.

(c) tetrad, (d) cube, (e) chain, and (f) clumps. The formation of tetrads and cubes is possible in spheres where it is not in the rods, as the rods always divide along the transverse axis. But in spheres there is no definite transverse axis. The fact that some of the spheres form definite patterns aids in their identification. The *Sarcina* form cubes, while *Streptococcus* form chains and *Staphylococcus* form clusters. Of these three genera *Staphylococcus* is not recognized in the sixth edition of "Bergey's Manual of Determinative Bacteriology" but has been classified under the genus *Micrococcus*.

The spiral form occurs primarily as individual cells and diplos. The diplo form is merely the intermediate form during growth and division.

Size

The size of the individual cells vary over rather wide limits. The cell size changes with time during growth and death. The limits of cell sizes range from 0.3 micron to 50 microns. The limits for the common bacteria are from 0.5 micron to 3.0 microns. The average rod is from 0.5 to 1.0 micron wide to 1.5 to 3.0 microns long. The average sphere is from 0.5 to 1.0 micron in diameter, while the average spiral cell is from 0.5 to 5 microns wide and 6 to 15 microns long.

Cell Structure

The cell structure can best be studied in the rod form. A schematic diagram of a typical bacterium is shown in Fig. 3-3. If the cell is ap-

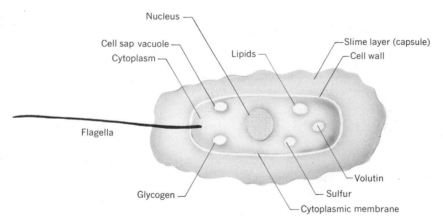

FIG. 3-3. Schematic diagram of a bacterium.

proached from the outside toward the center, the following components would be uncovered.

Slime Layer. The slime layer is the outermost layer of the bacteria. Although considerable information is known about the chemistry of the slime layer, little is known of its function. The slime layer is an accumulation of polysaccharide around the cell. It does not play any part in the growth of the cell, but appears to be a degradation product of cell-wall decomposition. In the early days it was thought that the slime layer was a secretion from within the cell but the size of the molecule was found to be too large to pass through the cell wall.

The slime layer has always been considered important in trickling filter slimes and in activated sludge floc. It was postulated that certain bacteria secreted special slime to cement the bacteria together and to attract food. This concept has been proved completely false. Slime formation is a normal result of metabolism by all bacteria. If the bacteria are actively motile, the slime is sheared from the ·cell surface. This phenomenon has caused some observers to wonder if all bacteria produce slime, as it is not visible on very young, very motile bacteria. Motility causes the slime layer to take the form of the cell and appear at a regular thickness around the cell. When the slime layer has a regular thickness, it is known as a *capsule*.

The ease with which the slime layer is washed from the cell surface indicates that the structural strength of the polysaccharide molecules is weak. Examination of the structure of the slime polysaccharide shows few reactive chemical groups, making this material chemically inert. This is further substantiated by efforts to demonstrate the slime layer by staining. The slime layer cannot be stained by normal acidic or basic dyes because there are few chemically reactive groups in the slime polysaccharide. The slime polysaccharide can be demonstrated only when treated with chemicals to oxidize the polysaccharide to yield reactive groups or by the use of a dye which reacts with the internal hydroxyl (^{-}OH) groups in the basic polysaccharide molecules. The inertness of the slime polysaccharide prevents its degradation by bacteria and allows it to accumulate.

When bacteria age and motility is reduced, the polysaccharide slime accumulates around the cell to a greater extent. Nonmotile cells accumulate so much slime that they have been designated as zoogloea forms. The presence of large quantities of slime around the bacteria indicates an old system of low metabolic activity.

Cell Wall. The cell wall lies below the slime layer and is responsible for the shape of the cell. Chemically, the cell wall is a protein-polysaccharide polymer of low chemical reactivity. The molecular structure of the cell wall is such that relatively large molecules are unable to pass through the cell wall. In this way the cell wall acts as a sieve to control the size of molecules entering the cell.

Cytoplasmic Membrane. The cytoplasmic membrane lies below the cell wall. It is a lipoprotein complex which controls the passage of all materials into and out of the cell. The presence of many highly reactive chemical groups guides the incoming materials to the proper points within the cell for further reaction. Control of materials into the cell is accomplished by physical screening, as well as by electric charge. The cytoplasmic membrane is the site of the surface charge of the bacteria.

Cytoplasm. The mass of the bacterial cell is made up of a colloidal suspension of proteins, lipids, and carbohydrates known as cytoplasm. The major chemical reactions occur within the cytoplasm at the colloidal surfaces. The aqueous phase is a mobile phase in which the soluble chemicals move. The end products of metabolism are released from the colloidal surface into the aqueous phase and then out of the cell. There are a number of major insoluble structures within the cytoplasm, the chief of which is the nucleus.

Nucleus. The heart of the bacterium is the nucleus. It is composed of nucleoproteins and is responsible for all chemical reactions which occur within the cell. The enzymes which catalyze all biochemical reactions within the cell have their origin in the nucleus. The hereditary characteristics of the bacteria stem from the nucleus. As long as the nucleus remains intact, the cell can continue to function, as it still retains the ability to repair damage and to create new cellular components.

Inclusions. Other inclusions within the cytoplasm vary from cell to cell and vary with certain environmental limits. The more common inclusions are volutin, polysaccharide, lipid, and sulfur. Volutin is a pentosenucleic acid which can be used as a reserve for the synthesis of the nucleus. Polysaccharide inclusions are either glycogen or starch and are found in the presence of excess organic matter or in a deficit of nitrogen. Lipid or fatty inclusions appear when there is an excess of high-energy food such as carbohydrates. Both the polysaccharide and lipid inclusions act as food reserves when needed for energy metabolism. The sulfur-metabolizing bacteria often accumulate free sulfur within the cell. None of these inclusions occur in bacteria except under conditions of excess energy nutrients and a deficit of synthesis nutrients.

Flagella. The majority of bacteria of interest to sanitary engineers are free swimming. The organs associated with motility have been termed flagella. There has been a definite controversy as to the mechanism of motility. Some investigators felt that flagella were not the cause of motility but were artifacts caused by streaming polysaccharide slime. Recent studies have shown that the flagella are basically protein, removing the possibility that flagella are polysaccharide artifacts. Electron photomicrographs have shown that the flagella have their origin within the cell and are definite structures.

Spores. The need of microorganisms to survive under adverse environmental conditions has resulted in the formation of spores by some species of bacteria. Bacterial spores result only when normal, healthy cells find themselves in a slowly changing adverse environment. The spore results when the nucleus becomes surrounded by a very tough polysaccharide coating. The polysaccharide coating protects the nucleus until favorable environmental conditions are established, at which time the nucleus expands to a full cell and the polysaccharide coating is discarded.

Chemical Composition

The normal growth of bacteria in excess nutrients results in a bacteria cell of a definite chemical composition. Examination of many different bacteria grown under many different environmental conditions indicates that bacteria are 80 per cent water and 20 per cent dry matter. The dry matter is 90 per cent organic and 10 per cent inorganic. The organic fraction consists of 53 per cent carbon, 29 per cent oxygen, 12 per cent nitrogen, and 6 per cent hydrogen. The organic fraction gives an approximate empirical formulation of $C_5H_7O_2N$. The inorganic fraction averages 50 per cent P_2O_5, 6 per cent K_2O, 11 per cent Na_2O, 8 per cent MgO, 9 per cent CaO, 15 per cent SO_3, and 1 per cent Fe_2O_3. The bacteria must derive all the basic elements for protoplasm from the liquid environment. If the environment is deficient in one or two elements, the bacteria will develop only in proportion to the chemical deficiency.

Metabolism

Metabolism determines the bacteria's ability to grow in any environment. Not only must the environment supply the chemical elements to produce the protoplasm for new cells but it must also furnish sufficient energy to permit the bacteria to synthesize the protoplasm. Bacteria obtain their energy either from sunlight or from oxidation of chemical compounds.

Photosynthesis. A few bacteria utilize the energy from sunlight to synthesize protoplasm. The energy from sunlight is converted by photosynthesizing pigments similar to chlorophyll found in higher plants. The use of energy from sunlight permits the bacteria to utilize carbon dioxide as their source of carbon for protoplasm. The mechanism of carbon dioxide fixation by photosynthetic bacteria differs from that of higher plants in that oxygen is not evolved. The basic equations for photosynthesis by bacteria and higher plants are shown in Eqs. (3-1) and (3-2).

Higher Plants

$$XCO_2 + NH_3 + ZH_2O \xrightarrow{\text{Sunlight}} C_xH_yO_zN + XO_2 \tag{3-1}$$

Bacteria

$$XCO_2 + NH_3 + \frac{x}{2}H_2S + ZH_2O \xrightarrow{\text{Sunlight}} C_xH_yO_zN + \frac{x}{2}H_2SO_4 \tag{3-2}$$

Basically, the two equations are similar, with the exception that the bac-

terial photosynthesis reaction oxidizes hydrogen sulfide rather than releasing free oxygen. There is some evidence that the blue-green algae
will react in the same manner as the photosynthetic bacteria if hydrogen
sulfide is present. Very little is known about these bacteria and they are
of little importance to sanitary microbiologists other than to know that
photosynthetic bacteria do exist.

Chemosynthesis. The oxidation of inorganic or organic compounds to
obtain energy for synthesis is called chemosynthesis and is the most common method of bacterial metabolism. The bacteria which oxidize inorganic compounds utilize carbon dioxide for their protoplasm and are
called autotrophic bacteria. The bacteria which oxidize organic compounds for energy obtain the carbon for synthesis from the same organic
compound used for energy. A portion of the organic compound is used
for energy and a portion is used for synthesis. These bacteria are known
as heterotrophic. Both the autotrophic and the heterotrophic bacteria
are important to sanitary microbiology.

Autotrophic Bacteria. The autotrophic bacteria are the most complex
group of bacteria from a bio-
chemical standpoint. These bacteria have the ability to make all
the complex chemical structures
within the bacterial cell from the
basic inorganic chemicals in water. The autotrophic bacteria have
not been studied extensively because of their limited importance
to bacteriologists. They have been
studied primarily as freaks. It suffices to say that these bacteria are
all strict aerobes, utilize carbon
dioxide as their carbon source, and
are classified by their energy
source. Certain of the autotrophic
bacteria are extremely important
to the sanitary microbiologist in
waste treatment plants and in corrosion. It is expected that interest
in the biochemistry of the autotrophic bacteria will increase as
interest in sanitary microbiology
increases.

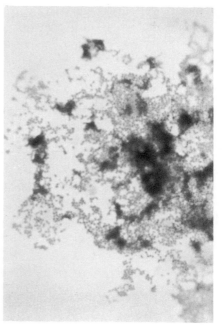

FIG. 3-4. Photomicrograph of bacteria
(4,000 ×).

Heterotrophic Bacteria. The heterotrophic bacteria are the most important group of bacteria. They require organic compounds which sup-

ply their carbon and their energy. The heterotrophic bacteria can be broken into three groups based on their relationship with oxygen in the energy reaction. The heterotrophs which utilize free dissolved oxygen are known as aerobes. The heterotrophs which oxidize organic matter in the complete absence of dissolved oxygen are known as anaerobes. There is a group of bacteria which utilize free oxygen when it is present but which can also carry on metabolism in the absence of free oxygen. This latter group of bacteria is known as facultative. The facultative bacteria have sometimes been designated as facultative aerobes or facultative anaerobes. All three terms are synonymous.

Classification

The varied forms which the bacteria take and the varied mechanisms of synthesis and oxidation have made the problem of bacterial classification difficult and confusing. To say that bacterial taxonomy is confusing is an understatement of the highest magnitude. The greatest frustration facing the sanitary microbiologist lies in bacterial taxonomy. Yet, it is not hopeless. The Society of American Bacteriologists has taken on the problem of bacterial taxonomy and is making slow but steady progress. Their biggest handicap is that they started so late.

The Society of American Bacteriologists has set forth recommendations for the tests to be made to determine the identity of unknown bacteria. Anyone interested in bacterial identification should follow the procedures outlined in the "Manual of Microbiological Methods."

Bacterial classification is based on form, stain reactions, and biochemical reactions. The more tests that are run, the more differentiation that can be made. Needless to say, there is a practical limit to the number of tests run on any pure culture bacteria. This is where judgment is required and judgment comes best by experience. Once the biochemical characteristics have been determined, the bacteria are identified from descriptions published in "Bergey's Manual of Determinative Bacteriology" which is also published under the auspices of the Society of American Bacteriologists.

SUGGESTED REFERENCES

1. Knaysi, G., "Elements of Bacterial Cytology," 2d ed., Comstock Publishing Associates, Inc., Ithaca, N.Y., 1951.
2. Society of American Bacteriologists, "Manual of Microbiological Methods," H. J. Conn (ed.), McGraw-Hill Book Company, Inc., New York, 1957.
3. "Bergey's Manual of Determinative Bacteriology," 7th ed., The Williams & Wilkins Company, Baltimore, Md., 1957.

CHAPTER 4

Fungi

Fungi are very similar to bacteria. In fact, to be technical, the bacteria are actually fungi, fission fungi. The importance of bacteria has caused them to be singled out and studied separately from the fungi. This fact has caused common usage of the term fungi to mean all fungi except bacteria. Two other terms which require consideration are mold and yeast. Mold is synonymous in common usage with fungi. Yeast are a part of the fungi but like the bacteria, their importance has caused a separation from fungi similar to the bacteria. This division of terminology has led to some confusion in understanding the fungi by sanitary engineers.

Definition

By definition fungi include the nonphotosynthetic plants and if we wish to exclude the bacteria we can add the requirement of being multicellular. The absence of photosynthetic pigments requires the fungi to utilize organic matter as their source of carbon and energy. It is this property of metabolism of organic matter that makes both the fungi and the bacteria important to the sanitary engineer. The fungi have two major advantages over the bacteria in that they can grow in low-moisture areas and can grow in low pH solutions. For these reasons the fungi play a major role in composting organic matter and in the treatment of some industrial wastes.

Classification

The fungi are divided into five classes:
1. Myxomycetes—slime fungi
2. Phycomycetes—algae fungi
3. Ascomycetes—sac fungi
4. Basidiomycetes—rusts, smuts, and mushrooms
5. Fungi Imperfecti—miscellaneous

Of these five classes the sanitary engineer is primarily interested in three, Phycomycetes, Ascomycetes, and Fungi Imperfecti. Theoretically, the Myxomycetes should also be of interest; but so little is known of this class that they are not considered important at the present time.

The relatively few classes of fungi leads the sanitary engineer often to oversimplify the problems of classification of fungi. Actually, there are

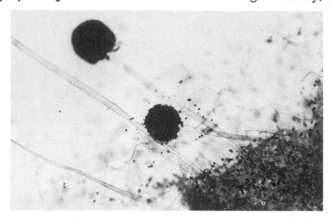

FIG. 4-1. Photomicrograph of the fungi, *Aspergillus,* showing sporangium and spores (4,000 ×).

1,500 different species of Phycomycetes, between 25,000 and 35,000 species of Ascomycetes, and from 15,000 to 20,000 species of Fungi Imperfecti. In view of the fact that there are approximately 50,000 known species of fungi in these three classes, the problem of classification becomes an absurdly complex phenomenon. Fortunately, the sanitary engineer has little interest or need for detailed study of fungi classification.

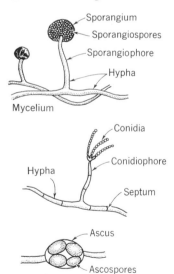

FIG. 4-2. Nomenclature of fungi.

Identification

Although the sanitary engineer is not interested in detailed classification of the fungi, he is interested in identification of fungi and groupings at least as to class. Unlike the bacteria the fungi are not identified according to their biochemical reactions but rather are identified by their physical characteristics. Since the fungi have several phases of their life cycle, the observation of all phases is not always an easy task. Once again though, the sanitary engineer is not interested in this detailed look at the individual fungus. Only a moderate attempt at crude identification of most fungi is necessary or even practical.

Terminology

Every branch of science has its own language for communication and mycology is no different. The terminology is as different from that of bacteriology as that of civil engineering is from chemical engineering. In order to work even at the edge of mycology, it is necessary for the student to acquire a slight vocabulary as to the parts of the fungi.

DEFINITION OF TERMS

1. *Spore*—reproductive stage of the fungi
2. *Hypha*—a single filament (pl. hyphae)
3. *Septate*—a transverse wall across the hypha
4. *Nonseptate*—transverse wall across the hypha absent
5. *Mycelium*—the mass of hyphae
6. *Vegetative mycelium*—the mycelium which is responsible for absorption of food
7. *Reproductive mycelium*—the mycelium which is responsible for spore formation
8. *Sporangiophore*—the spore-producing structure in the Phycomycetes
9. *Sporangium*—the sac structure at the end of the sporangiophore
10. *Sporangiospores*—the spores within the sporangium
11. *Conidiophore*—the asexual spore-producing structure in the Ascomycetes
12. *Conidium*—the spore on the conidiophore
13. *Budding*—the process of reproduction in yeast by swelling
14. *Blastospores*—spores formed by budding
15. *Chlamydospore*—a spore formed by a resting cell by swelling and increasing the cell-wall thickness
16. *Arthrospores*—spores formed by fragmentation of the hyphae
17. *Ascus*—the sac structure containing the sexual spores of the Ascomycetes
18. *Ascospore*—the sexual spores contained in the ascus

Microscopic Examination

Microscopic examination is the key to identification of the fungi. For this reason it is important for the student to become proficient in the use of the microscope and in the recognition of what he sees under the microscope. The fungi can be examined directly or suspended in liquid, dried, and stained. The fact that the fungi are quite large, 5 to 10 microns wide, makes it easy to distinguish them from the filamentous bacteria or actinomycetes. The fungi do not contain as much protein as bacteria and, hence, stain lighter with the normal bacteriological dyes. True branching of the hyphae is another characteristic used to identify the fungi (Fig. 4-3).

(*a*) True branching   (*b*) False branching

FIG. 4-3. Schematic sketch of true and false branching.

Cultivation

One of the easiest ways to determine the presence of fungi is to grow them on culture media. Fungi can easily be detected by their rapid growth and aerial reproductive mycelium formation on Sabourauds glucose agar or on mycological agar. Commercial media are available from Difco or Baltimore Biological Laboratory for general growth of fungi.

The mycological media depend on the fact that fungi grow very easily on a high carbohydrate medium at pH 4.5 or can grow in the presence of antibiotics at pH 7.0. By depressing the pH of the culture medium to 4.5 it is possible to prevent most bacteria from growing, leaving only the

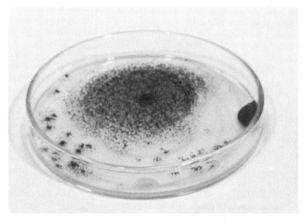

FIG. 4-4. *Aspergillus niger* growing on surface of agar medium with black aerial mycelium.

fungi to grow. The use of antibiotics to control extraneous bacteria has allowed cultivation of fungi at normal pH levels.

The isolation of fungi in pure culture requires the use of solid media in the same manner as bacteria. Fungi grow so poorly on liquid medium that it is not used very often. Yet growth on liquid medium offers a good chance to examine the vegetative portion of the fungi which is normally below the agar surface.

Aerobes

Fungi are aerobic organisms. When grown on a liquid medium, the fungi grow only at the liquid-air interface and do not diffuse through the initial medium as do facultative bacteria. The aerobic nature of fungi is of extreme importance to the sanitary engineer since it means the fungi will not be important in anaerobic digestion but rather only in aerobic systems.

There has been confusion as to the strict aerobic nature of fungi. This has stemmed from the fact that fungi produce end products typical of anaerobic metabolism. The problem lies in the quantity of oxygen available. If there is not sufficient oxygen to metabolize the organic matter completely to carbon dioxide and water, a portion of the organic matter will be oxidized to carbon dioxide and water while a portion will be oxidized part way. This phenomenon creates the impression of anaerobic metabolism but is still aerobic.

Reproduction

Fungi reproduce by spore formation. In bacteria spore formation is a survival mechanism and not a reproductive mechanism. This is a key difference between bacteria and fungi. Most of the spore formation is asexual but there is some sexual spore formation in the Ascomycetes. As indicated in Fig. 4-4, the spores are produced primarily in the aerial mycelium. This permits the spores to be easily spread by wind currents.

The fungi spore is similar in structure to the bacteria spore. It consists of a tough polysaccharide coating which protects the nucleus from prolonged desiccation. When the spore arrives at a suitable food source, germination begins and the fungi structure is produced, ready to reproduce the cycle again. Fungi produce tremendous numbers of spores. Fortunately, most fungi are not pathogenic to man. The lack of suitable environment prevents most fungi spores from ever germinating.

Fungi spores are transmitted great distances in the air by wind currents. Examination of air several miles above the earth has revealed the presence of fungi. Needless to say, fungi are the chief contaminants of bacteriological media in the laboratory.

The presence of fungi spores in digesting sewage sludge has raised the question of what the fungi were doing in the digester. The answer is simple, nothing. Fungi spores have the ability to survive in unfavorable environments for long periods of time. When plated on the proper media, they will grow. Strict aerobic bacteria and fungi can be isolated from anaerobic materials because of their spore form, but it does not mean that these microorganisms are metabolically active in the anaerobic environment.

Chemical Composition

Like the bacteria, the fungi produce definite chemical structures which make up their protoplasm. The quantity of information on fungi is far less than for bacteria. This reflects on the lack of interest in biochemistry by the mycologists. The fungi contain 75 to 80 per cent water, as do bacteria. An analysis of *Aspergillus niger* showed 47.9 per cent carbon, 6.7 per cent hydrogen, and 5.24 per cent nitrogen. The organic fraction gives an approximate empirical formulation of $C_{10}H_{17}O_6N$. Comparing the formulation of fungi and bacteria shows the immediate difference in nitrogen content. The fungi form normal protoplasm with one-half the nitrogen required by the bacteria. Thus, it is not surprising that fungi predominate over bacteria in nitrogen-deficient environments. A nitrogen-deficient environment for the bacteria is not nearly so deficient for the fungi. Thus it is that the chemical composition of the cellular material assists the sanitary microbiologists in understanding growth of the different groups of microorganisms.

Mycology—Science or Art

It becomes readily apparent that mycology is very complex, too complex. The problem of identification and classification of fungi has become so complex that most mycologists have little time for studying the biochemistry of the fungi they identify. This has led to a scarcity of information with regard to their biochemistry and has frightened most microbiology students away from mycology. It has definitely prevented any data from being obtained on the sanitary significance of fungi. Recognizing the importance of fungi in sanitary engineering is the first step. Other steps will soon follow.

SUGGESTED REFERENCES

1. Foster, J. W., "Chemical Activities of Fungi," Academic Press, Inc., New York, 1949.
2. Wolf, F. A., and F. T. Wolf, "The Fungi," 2 vols., John Wiley & Sons, Inc., New York, 1947.
3. Cochrane, V. W., "Physiology of Fungi," John Wiley & Sons, Inc., New York, 1958.

CHAPTER 5

Algae

The third group of microscopic aquatic plants are the algae. The algae differ from the fungi and bacteria in their ability to carry out photosynthesis. The algae can utilize the energy in light and do not have to depend upon the oxidation of matter to survive. In fact, the algae evolve oxygen during their growth.

The evolution of oxygen by algae and their production of taste-producing oils have made the algae of extreme interest to the sanitary microbiologists. The oxygen production of algae has been coupled with the oxygen demand of bacteria in the stabilization of sewage in oxidation ponds. On the other hand, the excess growth and death of some algae in water reservoirs has caused the sanitary microbiologist many problems in tastes and odor control.

Definition

There is no clear-cut definition of algae which can satisfy everyone. The simplest definition of algae is that it includes all microscopic plants carrying out true photosynthesis. In this way the photosynthetic bacteria which oxidize hydrogen sulfide are excluded. As will be seen later, this definition is naïvely oversimplified.

Identification

The presence of photosynthetic pigments makes it very easy to identify algae under the microscope. All identification of algae, and hence classification, is based upon physical characteristics. Unlike fungi, a single observation is normally sufficient for identification.

Classification

The classification of algae is constantly changing, with some phytologists preferring five divisions, some seven divisions, and some nine divisions. The classification used in Smith's "Fresh Water Algae of the United States" has seven divisions and will be used herein.

1. Chlorophyta—green
2. Euglenophyta—motile green

41

3. Chrysophyta—yellow-green to golden brown

4. Pyrrophyta—motile greenish tan to golden brown

5. Cyanophyta—blue-green

6. Phaeophyta—brown (marine)

7. Rhodophyta—red (marine)

It can readily be seen that if the two groups of marine algae, Phaeophyta and Rhodophyta, were dropped there would be only five divisions. This is the situation for fresh-water algae.

FIG. 5-1*a*. Photomicrograph of *Euglena* (4,-000 ×).

Observation

Algae are observed primarily in the liquid state.

It is very difficult to obtain satisfactory dried and stained specimens of algae. The hanging-drop slide is used for most simple microscopic examinations. For quantative examinations the Sedgwick-Rafter counting cell is used. The Sedgwick-Rafter counting cell holds exactly 1 ml of liquid in a chamber 20 by 50 by 1 mm. Most observations are made at 50 × or 100 × magnification.

Being living material, algae will die if kept in storage for more than a short period of time. If samples are not to be observed within a few hours of collection, it is necessary to add a preservative. The most common preservative is formaldehyde. Sufficient formaldehyde to make a 2 per cent solution with the sample will preserve the algae indefinitely. The only problem is that the formaldehyde-preserved algae will not retain their pigment indefinitely.

Normally, algae of sanitary significance are sufficiently concentrated that observations can be made on the sample directly. In many instances the sample will have to be diluted to give sufficient space to examine the organisms. In a few instances the con-

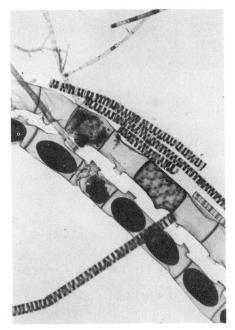

FIG. 5-1*b*. Photomicrograph of *Spirogyra* (4,000 ×).

centration of algae in the sample will be too low for normal examination. In this instance it will be necessary to concentrate the algae. The easiest way to concentrate the microorganisms is with a centrifuge. A small laboratory centrifuge will concentrate the organisms in a small sample sufficiently for examination. The Sedgwick-Rafter filter is still used by phytologists, but the author has never seen a need for the Sedgwick-Rafter filter in normal sanitary analyses.

FIG. 5-1c. Photomicrograph of *Asterionella* (4,000 ×).

The newest technique for concentrating and counting the algae is the use of the membrane filter. With dilute algae samples a relatively large volume of liquid can be passed through the membrane filter. The algae are retained on the filter surface and can be counted under the microscope. If the algae are too concentrated, they will overlap and be impossible to count.

Pure Cultures

Unlike bacteria and fungi it is not possible to isolate and to grow all types of algae in pure culture. Only a few of the simpler forms of algae can be isolated and grown in pure culture. Simple single-celled forms such as *Chlorella* can be isolated in the same manner as bacteria on tryptone glucose agar. Some of the higher forms of algae can be washed free of extraneous microorganisms and then grown in a suitable liquid medium.

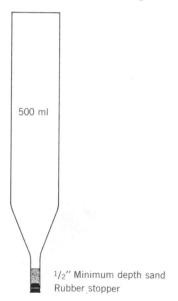

500 ml

1/2″ Minimum depth sand
Rubber stopper

FIG. 5-2. Sketch of Sedgwick-Rafter filter.

Culture Media

Algae are autotrophic organisms in that they utilize inorganic compounds for their protoplasm. The mineral requirements for algae protoplasm is similar to that of bacteria protoplasm. Carbon comes from CO_2 as it does for the autotrophic bacteria. Nitrogen can be used in the form of ammonia, nitrite, or nitrate. Phosphorus is always in the orthophosphate state, while sulfur is as the sulfate. The normal trace elements of sodium, potassium, calcium, magnesium, iron, cobalt, and molybdenum are all required. Thus it is that any culture media must be made up of all the common nutrient salts.

Each group of algae has its own special nutrient requirements.

This necessitates the use of different synthetic media for the special groups of algae. The Euglenophyta grow best in a rich ammonia nitrogen medium. Other algae such as the desmids prefer low pH or soft water, while others such as the Cyanophyta prefer high pH or hard water. The easiest way to prepare synthetic culture media is to make mineral analyses of the water in which the algae are known to grow well and to make the media accordingly.

Metabolism

The algae, like the other microorganisms, are concerned with reproduction of the species. All their metabolic activities are directed toward staying alive and producing new cells. In order to grow, it is necessary to have certain building blocks, as already indicated. The quantity of the various elements required can be estimated from an analysis of the algae. Unfortunately, the phytologists have not been interested in the chemical composition of algae and only a few analyses are available. An analysis of *Chlorella* gave an empirical formulation of protoplasm as $C_5H_8O_2N$ (Burlew). Another analysis (Fogg) of this same organism yielded $C_{5.7}H_{9.8}O_{2.3}N$.

In the presence of sunlight the algae convert the inorganic materials in the water into organic matter in the form of protoplasm. It has been erroneously assumed by some that the end product of photosynthesis is polysaccharide according to the following equation:

$$CO_2 + 2H_2O \xrightarrow{\text{Sunlight}} (CH_2O) + O_2 + H_2O \qquad \text{(Fogg)} \qquad (5\text{-}1)$$

The extra mole of water is shown on both sides of the equation since the oxygen evolved has been shown to come from the water and not the carbon dioxide. There is no doubt that the pattern of metabolism starts with the general equation above but the unit of carbohydrate (CH_2O) is not the end product. The over-all equation for the growth of algae can be expressed by the following equation obtained from *Chlorella* (Fogg):

$$NH_3 + 5.7CO_2 + 12.5H_2O \xrightarrow{\text{Sunlight}} C_{5.7}H_{9.8}O_{2.3}N + 6.25O_2 + 9.1H_2O \qquad (5\text{-}2)$$

Thus it is that cellular protoplasm is the end product of metabolism along with oxygen.

The ability of algae to produce oxygen makes it of extreme significance in sanitary engineering. Bacteria in conjunction with algae have been used in the oxidation pond method for sewage treatment. This process will be discussed in detail in a later chapter. But it should be recognized that algae liberate oxygen only while they are synthesizing organic matter.

In the absence of sunlight some of the algae are able to carry on chemosynthetic metabolism like bacteria. In doing so the algae require oxygen for oxidation. Other algae undergo endogenous metabolism with degradation of their own protoplasm to supply energy for survival. The endogenous metabolism equation is as follows:

$$C_{5.7}H_{9.8}O_{2.3}N + 6.25\ O_2 \rightarrow 5.7CO_2 + NH_3 + 3.4H_2O \qquad (5\text{-}3)$$

The demand by algae for oxygen in the absence of sunlight is of as much importance as their production of oxygen in its presence.

Pigments

The photosynthetic activity of algae is determined by the photosynthetic pigments which exist within the cells. The four major groups of photosynthetic pigments are:

1. Chlorophylls
2. Carotenes
3. Xanthophylls
4. Phycobilins

The Chlorophyta, grass-green algae, have predominantly chlorophyll-type pigments as do the Euglenophyta. The yellowish green and golden brown colors of the Chrysophyta are due to the predominance of carotenes and xanthophylls. The Phaeophyta are brown because of the excess xanthophyll pigments. The Pyrrophyta contain chlorophyll, carotenes, and xanthophylls, with the xanthophylls predominating. The blue-green Cyanophyta show the influence of the phycobilins along with the other three pigments. The Rhodophyta have predominantly phycobilin pigments.

SUGGESTED REFERENCES

1. Smith, G. M., "The Fresh-water Algae of the United States," 2d ed., McGraw-Hill Book Company, Inc., New York, 1950.
2. Burlew, J. S., Algal Culture, *Carnegie Inst. Washington Publ. No. 600*, Washington, D.C., 1953.
3. Fogg, G. E., "The Metabolism of Algae," John Wiley & Sons, Inc., New York, 1953.

CHAPTER 6

Protozoa and Higher Animals

Of equal importance with the microscopic plants are the microscopic animals. For the most part the microscopic animals are scavengers which tend to clean the excess bacteria from solution. This action is significant to the sanitary engineer in biological waste treatment systems in producing clarified effluents and in stimulating maximum bacterial growth. In streams and lakes the lower animals form an important link in the gross aquatic biological cycle. The microscopic animals eat the microscopic plants and in turn are eaten by higher animals.

PROTOZOA

Definition

It is not possible to define protozoa easily; in fact, it has been considered by some as impossible to define. For our purposes it suffices to

FIG. 6-1a. Photomicrograph of *Paramecium* (400 ×).

say that the protozoa are single-cell animals which reproduce by binary fission. Most of the protozoa metabolize solid food and have more complex digestive systems than the microscopic plants. As is the case with most classifications, there are transition forms which have the characteristics of both animals and plants. The transition forms are normally classified according to their predominant characteristics.

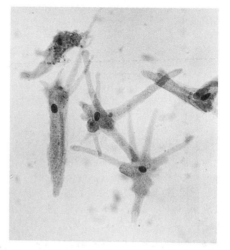

FIG. 6-1b. Photomicrograph of *Amoeba proteus* (400 ×).

Identification

Protozoa are easily recognized by their physical characteristics. Most of the common protozoa can be observed easily at 100 × magnification, ranging in size from 10 to 100 microns. Their size makes it easy to determine if the unknown animal has a single-cell body or a multicell body. All protozoa are motile although some are attached to solid particles. Some common protozoa are shown in Fig. 6-1. Protozoa are observed largely in their natural environment and are seldom dried and stained. The reason for this lies in the fact that the protozoa lose their shapes on drying and hence many important characteristics required for classification.

Classification

The classification of protozoa is almost as confusing as the definition of protozoa. One major reason for this lies in whether or not the motile, single-cell, photosynthetic organisms are algae or protozoa. The phytologists claim them as algae while the protozoologists claim them as protozoa. The microbiologist who is interested in both algae and protozoa is faced with the dilemma of what classification to use. The author has chosen to consider all the single-cell, photosynthetic organisms as algae. This makes it easier for the student who is already overburdened with taxonomy for taxonomy's sake.

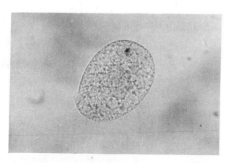

FIG. 6-1c. Photomicrograph of *Glaucoma* (4,000 ×).

Since the sanitary microbiologist is not interested in detailed taxonomy of protozoa and other microorganisms, the author presents only a grossly oversimplified classification which he has found of practical value in this field. The protozoa have been divided into five groups with their chief characteristic:

1. *Sarcodina*—pseudopodia (false foot)
2. *Mastigophora*—flagella
3. *Sporozoa*—parasitic sporeformer
4. *Ciliata*—cilia
5. *Suctoria*—tentacles

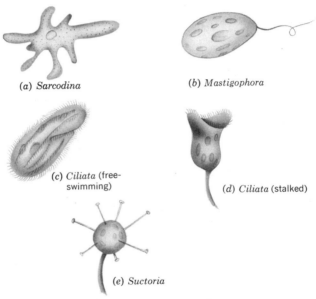

(a) *Sarcodina*

(b) *Mastigophora*

(c) *Ciliata* (free-swimming)

(d) *Ciliata* (stalked)

(e) *Suctoria*

Fig. 6-2. Five basic groups of protozoa.

Sarcodina. The Sarcodina are easily distinguished by their flowing protoplasm. The cells are without definite cell walls although some have rigid shells. Movement is primarily by pseudopodia, the false foot. The *Amoeba* is the most common Sarcodina. Almost everyone has observed the shifting of protoplasm by the *Amoeba* in moving. The Sarcodina do not play a major role in sanitary microbiology. Only the pathogen, *Endamoeba histolytica*, has received much attention. This organism will be discussed later with regard to water and sewage.

Mastigophora. The Mastigophora have definite cell walls and move by means of one or two flagella. The flagella are long whiplike organs which give the cell an irregular motion. It is in the Mastigophora that the confusion has arisen as to the question of algae or protozoa. The Mastigophora are divided into two subgroups, *Phytomastogophora* and *Zoomasti-*

gophora. The Phytomastigophora resemble algae but they lack photo-synthesis. They metabolize organic matter in the same way that the bacteria do, by diffusion of soluble food across the cell wall. The Zoo-mastigophora have gullets which permit them to take in solid food particles. As a result, the Zoomastigophora utilize bacteria, algae, and protozoa as their source of food.

Sporozoa. The Sporozoa are entirely parasitic and are not of interest except in the medical aspects of sanitary microbiology. The Sporozoa have complicated life cycles of which the spore stage is their chief char-acteristic. The most common Sporozoa of interest to us is *Plasmodium vivax*, the causative agent of malaria.

Ciliata. The Ciliata are the most important protozoa to the sanitary microbiologist in stream pollution and waste treatment systems. The Ciliata move by small hairlike cilia which appear attached to the cell in a regular manner. It has been convenient to divide the Ciliata into the free-swimming forms and the stalked forms. The free-swimming Ciliata move rapidly through the liquid, metabolizing solid organic matter as fast as they can take it in. This rapid movement consumes considerable en-ergy and is the cause of their rather large appetite. The *Paramecium* is the best example of a free-swimming Ciliata.

Some of the Ciliata are attached by a stalk to a particle of matter and use their cilia to bring the food to them. If the attached particle is small, the action of the cilia is sufficient to propel the stalked Ciliata through the liquid.

Suctoria. The Suctoria have two phases in their life cycle, an early ciliated, free-swimming stage and an adult stalked stage. The adult stage is identified by the presence of rigid tentacles. The tentacles are used to clamp on stray free-swimming protozoa which are then drawn into the cell. The role of the Suctoria in sanitary microbiology has never been fully studied, but the Suctoria can normally be found in activated sludge waste treatment systems.

Metabolism

Although there have been reports of protozoa growing under anaerobic conditions, none of the common protozoa of sanitary significance are known to be anaerobic. Protozoa have never been reported as normal inhabitants of anaerobic digesters. All the protozoa of interest to sanitary microbi-ology are strict aerobes. This greatly simplifies the relationship of the protozoa for the sanitary microbiologists.

The metabolic pattern within the protozoa is similar to the nonphoto-synthetic microorganisms. Energy is obtained from the oxidation of or-ganic matter, while the building blocks of new protoplasm are also ob-tained from the organic matter.

The Phytomastigophora take in soluble organic matter across their cell wall. Most of the other protozoa can do the same thing if the food concentration in solution is sufficiently high. The majority of protozoa ingest solid particles of food. The food is enclosed within a food vacuole where it undergoes degradation. The soluble organics diffuse into the cell from the food vacuole and the waste materials remaining within the food vacuole are discharged from the cell into the liquid.

The metabolic activities of the protozoa lead to the production of protoplasm. As in the case of the other microorganisms, the protoplasm has a definite chemical formulation. Like the mycologists and phytologists, the protozoologists have not spent too much time on the chemical structure of the cellular protoplasm of their favorite microorganisms. The best available data on the empirical formulation of protozoa protoplasm indicate $C_7H_{14}O_3N$.

Nutrition

Many of the protozoa are unable to produce all the necessary growth accessory substances and depend on the bacteria they metabolize to furnish these substances. Since bacteria differ in their production of these substances, the protozoa tend to be selective in which bacteria they eat. Pure culture studies with the ciliate, *Tetrahymena*, and different bacteria have shown wide variations in growth of the protozoa. With some bacteria the ciliate would not even grow. Even the Phytomastigophora are affected by the bacteria around them. It has been shown that the Phytomastigophora can be stimulated by excess vitamin B_{12} produced by certain bacteria.

As already stated, most protozoa can utilize soluble organic compounds for their food. This is true if the organic concentration is high, 5,000 to 10,000 mg liter. The soluble organic concentration of liquid wastes is usually too low to support any protozoan growth other than the Phytomastigophora. The protozoa survive in the dilute organic wastes by living off the bacteria. Understanding this method of nutrition for protozoa is very important for the sanitary microbiologist, especially when considering the relative position of protozoa in mixed microbial populations.

ROTIFERS

As the animal progresses to higher forms, it goes from a single-cell animal to a multicell animal. The rotifer is the simplest of the multicell animals. It derives its name from the apparent rotating motion of the two sets of cilia on its head. The movement of these cilia makes the rotifer appear to have two contrarotating wheels. The cilia have a dual purpose, motility and food catching. Normal movement of the cilia causes the rotifer to move swiftly through the liquid. In order to feed at a fixed spot, the rotifer has a forked tail which permits it to attach itself to a solid particle. The rotifer has the ability to pull in tremendous quantities of liquid in search of food.

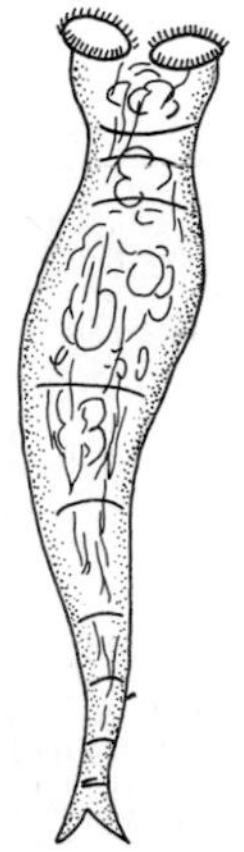

Metabolism

The rotifer is a strict aerobe and normally is found only when the environment contains at least several milligrams per liter dissolved oxygen. Bacteria form the chief source of food for the rotifers, but they can ingest any small organic particles. Because of their metabolic habits rotifers are found only in waters of low organic content. They act as good indicators of low pollutional waters.

CRUSTACEANS

Fig. 6-3. A sketch of a rotifer.

Normally, crustaceans bring to mind the hard-shelled animals such as the crayfish and the lobster. There are also microscopic crustaceans. The rigid shell structure is the chief characteristic of these multicellular microorganisms. Two of the most common crustaceans of interest to the sanitary engineer are *Daphnia* and *Cyclops*.

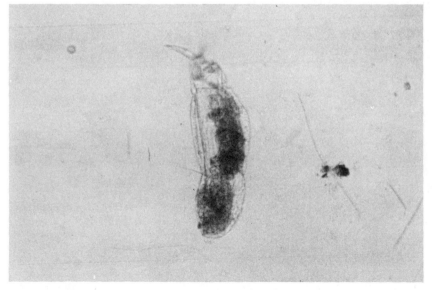

FIG. 6-4a. Photomicrograph of rotifer (400 ×).

The crustaceans are strict aerobes which feed on bacteria and algae. They are important as a source of food for fish. Crustaceans have also been used to clarify algae-laden effluents from oxidation ponds. The

FIG. 6-4b. Photomicrograph of the rotifer, *Anurea* (400 ×).

metabolic complexity of the crustaceans limits their growth to relatively stable streams and lakes.

WORMS AND LARVAE

Worms and larvae are normal inhabitants in organic muds and biological slimes. Nematodes have been found in activated sludge and in

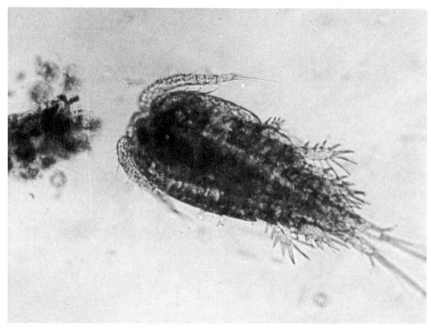

FIG. 6-5. Photomicrograph of the crustacean, *Cyclops* (400 ×).

trickling filter slimes. Little is known of their true function. They have definite aerobic requirements but can metabolize solid organic matter not readily degraded by the other microorganisms.

Two common organisms used in stream pollution studies as indicators of pollution are the worm, *Tubifix*, and the midge fly larvae, Chironomidae. Both of these organisms are red and are often confused by untrained observers. The *Tubifix* are normally found in very polluted streams while Chironomidae are found after the zone of active decomposition as the stream begins recovery. Other higher forms of animals are of passing interest to sanitary engineers and will be discussed in relation to their significance.

SUGGESTED REFERENCES

1. Lwoff, A., "Biochemistry and Physiology of Protozoa," vol. 1, Academic Press, Inc., New York, 1951.
2. Hutner, S. H., and Lwoff, A., "Biochemistry and Physiology of Protozoa," vol. 2, Academic Press, Inc., New York, 1955.
3. Hall, R. P., "Protozoology," Prentice-Hall, Inc., Englewood Cliffs, N.J., 1953.

CHAPTER 7

Enzymes

Microorganisms bring about many different chemical reactions, all of which are aided by organic catalysts known as enzymes. The purpose of the enzymes is to speed up the rate of hydrolysis of complex organic compounds and the rate of oxidation of simple compounds.

Nature of Enzymes

What are enzymes? Actually enzymes are complex proteins made up of three basic components: (1) apoenzyme, (2) coenzyme, and (3) metallic activator. The apoenzyme determines *where* the chemical reaction will occur. It is a protein of definite chemical structure. In fact, each enzyme has an entirely different apoenzyme. The apoenzyme is responsible for the strict specificity of enzyme reactions. Enzymes react with specific chemical compounds or groups and do not react randomly. For this reason the microorganisms must have the ability to produce all the varied chemical structures required for the apoenzymes. The specificity of the apoenzyme is determined by the physical arrangements of the amino acids making up the protein and the chemical structure of the compound undergoing reaction. Thus it is that knowledge of chemical structures is essential in understanding microbiological reactions.

The coenzyme is a separate part of the enzyme which determines *what* chemical reaction will occur. This is the reactive part of the enzyme. Like the apoenzymes the coenzymes are all definite chemical structures. Unlike the apoenzymes, many of the coenzymes have known chemical structures. The structure of three common coenzymes is shown in Fig. 7-1. Diphosphopyridine nucleotide (DPN) is responsible for hydrogen transfer. Adenosine diphosphonucleotide (ADP) is important in energy transfer reactions, while CoA is the key to molecular splitting. The apoenzyme is highly specific in its reaction but the coenzyme is not. The same coenzyme will tie up with different apoenzymes to produce the same chemical reaction on different chemical compounds. This multiplicity of roles by the coenzymes is of prime importance to the microorganisms.

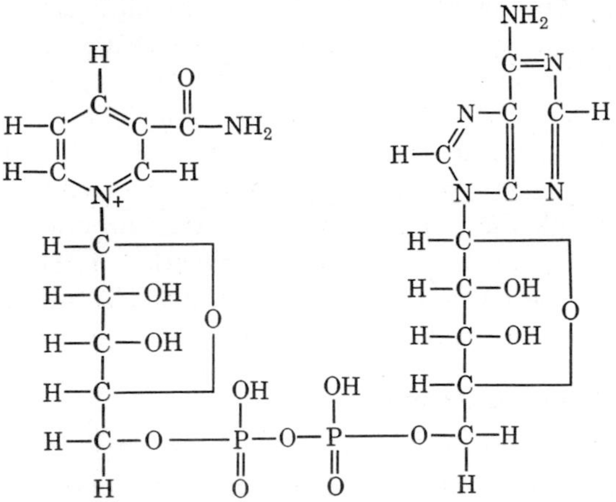

(a) Diphosphopyridine nucleotide (DPN)

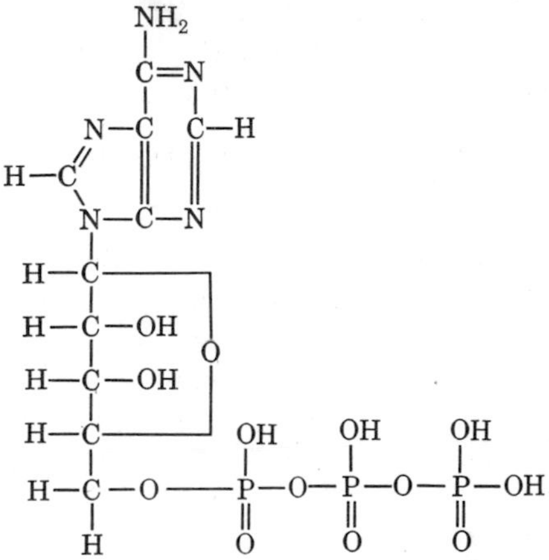

(b) Adenosine triphosphate (ATP)

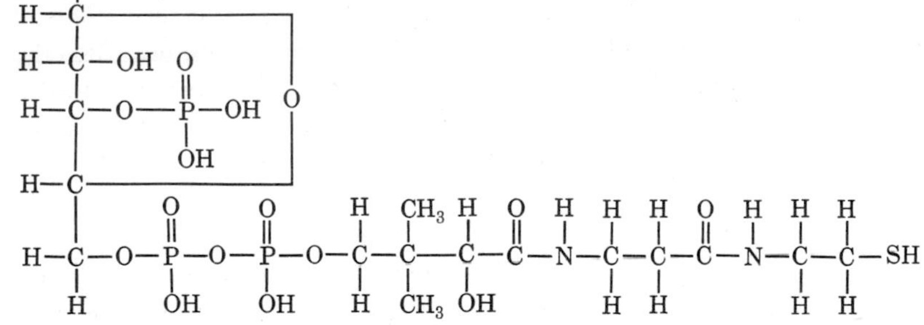

(c) Coenzyme A (CoA)

FIG. 7-1. Chemical structure of three common coenzymes.

The metallic activators are metallic cations which act as directors in lining up the enzyme and the substrate for perfect fit. The common metallic activators include potassium, calcium, magnesium, cobalt, iron, zinc, manganese, and molybdenum. The importance of perfect fit between the enzyme and the chemical material undergoing reaction cannot be overemphasized. A slight variation in chemical structure can completely block the reaction. Much of the direction of fit between enzyme and substrate is caused by the electric charges on various chemical groups in the enzyme and the substrate. The metallic activators bring the key parts of the reactants together in the exact manner required for reaction.

Mode of Action

Enzyme reactions are all reversible chemical reactions which are driven in accordance with the decrease in energy. No enzyme reaction occurs with a net increase in energy, as this is a fundamental impossibility. Being reversible, enzyme reactions are concentration-dependent. The concentration of reactant, the concentration of enzyme, and the concentration of end product all affect the reaction.

The basic enzyme reaction is

$$\text{Reactant} + \text{enzyme} \rightleftharpoons \text{reactant-enzyme complex} \rightleftharpoons$$
$$\text{enzyme} + \text{end product} \qquad (7\text{-}1)$$

Note that the enzyme is regenerated, while the reactant is changed to the end product. This is an essential feature of enzymes. Without the enzymes' continual regeneration, the microorganisms would spend all their efforts just to make new enzymes.

Equation (7-1) is an oversimplification of the complex enzyme reactions occurring within the microorganisms. A closer representation of the

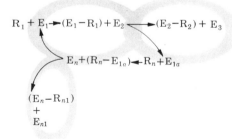

FIG. 7-2. Schematic diagram of enzyme reactions in coupled series.

enzyme reactions which occur within the cells is shown in Fig. 7-2. The organic substrate (R_1) undergoes reaction with the initial enzyme (E_1) to form the substrate-enzyme complex (R_1-E_1). The enzyme (E_1) brings about its chemical reaction and produces the end product (R_2). But the end product (R_2) is not released as a free compound. (R_2) is transferred directly from the (R_1-E_1) complex to the next enzyme (E_2) to form the (R_2-E_2) complex. Actually, the initial enzyme has either given up or obtained something from the substrate (R_1) so that the initial enzyme is not regenerated directly but remains in a modified state (E_{1a}). It is essential that (E_{1a}) be regenerated back to (E_1) if the chemical reactions are to continue. Thus it is that (E_{1a}) either gives up or receives from another reactant in the same series or, possibly, another series of reactants that portion of the molecule required to form (E_1) fresh for further reactions. The reactions proceed until the final end products are formed and released from the microorganisms.

Thus it is that the enzyme reactions within the bacteria are all interrelated by some series or scheme. It is impossible to look at a simple reaction correctly or completely without looking at its context in the entire microorganism. For any one step of the enzyme reaction it can be seen that the reaction is dependent upon the substrate concentration (R_1), the first enzyme concentration (E_1), the substrate-enzyme com-

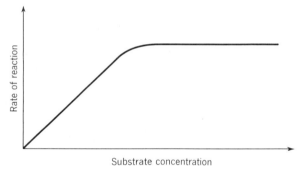

Fig. 7-3. Effect of substrate concentration on rate of reaction.

plex $(R_1\text{-}E_1)$, the modified first enzyme concentration (E_{1a}), and the regenerant concentration (Ri).

The rate of the enzyme reaction is increased by increasing the substrate concentration (R_1) or the enzyme concentration (E_1) within definite limits. Actually if the substrate concentration (R_1) is increased beyond a critical point, as shown in Fig. 7-3, the rate of the reaction is not increased proportionally. The enzyme concentration (E_1) has become the limiting factor. Increased second enzyme concentration (E_2) and the regenerant concentration (Ri) can increase the rate of reaction by removing the end product of the first reaction $(E_1\text{-}R_1)$ and by regenerating the enzyme $(E_{1a} \rightarrow E_1)$. A build-up of $(E_1\text{-}R_1)$ or (E_{1a}) will decrease the rate of the reaction since they act as a back pressure on the reaction.

In the series of enzyme reactions which occur within the microorganism the rate of the over-all reaction will be limited by the rate of the slowest step. A good example of this can be illustrated by the biological oxidation of methanol. If the methanol concentration is in excess, the limiting step of the reaction is in the oxidation of formic acid, as shown

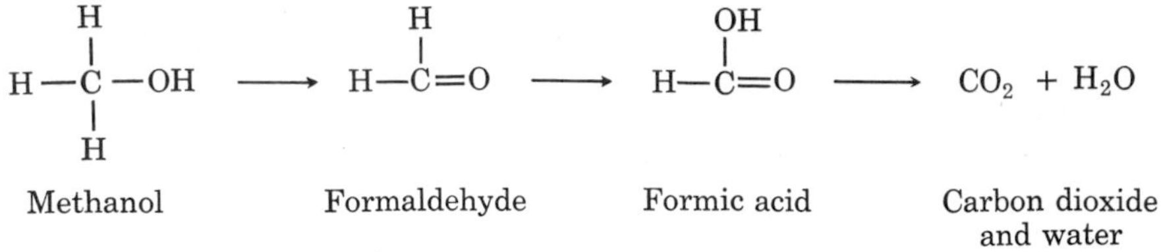

$$\underset{\text{Methanol}}{\overset{\displaystyle H}{\underset{\displaystyle H}{H-\overset{|}{\underset{|}{C}}-OH}}} \longrightarrow \underset{\text{Formaldehyde}}{\overset{\displaystyle H}{H-\overset{|}{C}=O}} \longrightarrow \underset{\text{Formic acid}}{\overset{\displaystyle OH}{H-\overset{|}{C}=O}} \longrightarrow \underset{\substack{\text{Carbon dioxide} \\ \text{and water}}}{CO_2 + H_2O}$$

Fig. 7-4. Metabolic reactions of methanol.

in Fig. 7-4. Formic acid will build up to the extent that it will pass from the microorganism into the surrounding environment. The normal limiting factor in the series reactions will be the regeneration of the critical enzyme.

Temperature and pH

There are two physical factors which have a very pronounced effect on enzyme reactions, temperature and the hydrogen-ion concentration, pH. The rate of enzyme reactions has been shown to approximately double with each 10°C rise in temperature up to 35°C. Above 35°C the protein fraction of the enzyme undergoes denaturation, resulting in the destruction of the enzyme. The effect of temperature on enzyme reactions is diagrammatically illustrated in Fig. 7-5.

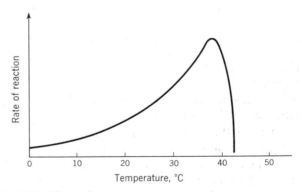

Fig. 7-5. Effect of temperature on rate of enzyme reactions.

There are some enzymes which do not undergo denaturation at 35 to 40°C but persist to the range of 65 to 70°C. These enzymes are part of the thermophilic or heat-resisting microorganisms. At the present time, little is known about the structure of the thermophilic enzymes and why they are heat-resistant. Much more research is needed to determine how the thermophilic enzymes differ from the normal or mesophilic enzymes.

The hydrogen-ion concentration, pH, exerts a pronounced effect on enzyme reactions, as shown in Fig. 7-6. The effect of pH differs with the particular enzyme in question. Some enzymes are optimum at low pH, while others are optimum at high pH. The decreased activity of the enzymes at low or high pH is due to two factors: (1) dissociation of key groups and (2) denaturation of proteins.

Dissociation. The enzyme proteins are made up of amino acids, as well as other components. The amino acids have two major reactive groups, amino (—NH_2) and carboxyl (—COOH). In neutral solutions the net positive electric charge on the amino groups equals the net negative electric charge on the carboxyl groups so that the net electric charge on the molecule is zero. As the pH decreases, there is an increase in hydrogen-ion concentration which tends to build up a greater positive charge on the amino groups (—NH_3^+). The molecule takes on a net positive electric charge at low pH. Increasing the pH above neu-

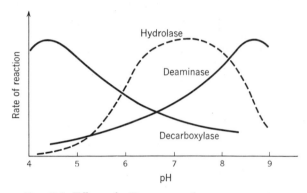

Fig. 7-6. Effect of pH on rate of enzyme reactions.

tral causes more carboxyl groups to undergo ionization (—COOH $\rightleftharpoons$ —COO^- + H^+). The net electric charge becomes negative. The changes in the electric charge of the enzymes as pH changes have a pronounced effect on the attraction or repulsion of the enzymes toward other molecules. At high pH the enzyme reactions are active on amino groups with attraction of the negatively charged enzyme to the positively charged amino groups. At low pH, the opposite is true.

Denaturation. High hydrogen-ion or hydroxyl-ion concentrations tend to denature the proteins in enzymes. Most enzymes begin to break down at pH 3 to 4, but the sulfate oxidizing bacteria appear to have enzymes capable of withstanding pH levels as low as 1.0. Once again little is known of the structure of these acid-resisting bacteria. Hydroxyl ions do not become significant until the pH reaches 9.5 or above. Algae have enzymes which operate at pH levels between 10 and 11 but this is not normal for other groups of microorganisms.

Salts and Heavy Metals

Neutral salts such as ammonium chloride, sodium chloride, sodium sulfate, etc., are used to precipitate proteins from solution by changing their solubility. Very high salt concentrations are required for this reaction, 10 to 30 per cent concentration. Heavy metals such as copper, zinc, calcium, magnesium, iron, etc., can also precipitate proteins. The heavy metals can tie up reactive points on the enzymes at concentrations below that required for precipitation. The use of heavy metals to inhibit enzyme activity is very important in sanitary microbiology.

Colloidal Nature

Enzymes are very large molecules which are not true soluble molecules but rather are colloids. It is believed that the large surface activity of the colloids is responsible for the enzyme properties. The enzymes within the microorganism exist in a colloidal suspension, thereby preventing their diffusion from the cell and subsequent loss. Bacteria and fungi must hydrolyze many external organic compounds before they can to be taken into the cell. The hydrolysis of organic matter outside the cell is brought about by enzymes located on the cell surface. As the cell ages, the bacterial surface decomposes and is renewed from within. This results in the release of the hydrolytic enzymes into solution and gives the appearance of the cell excreting enzymes into solution.

Extracellular and Intracellular Enzymes

The enzymes located on the cell surface are known as extracellular enzymes since they exist outside of the cell proper. The extracellular enzymes exist in a highly organized manner. Thus it is that the extracellular reactions occur at definite points on the cell surface. These reactions are all hydrolytic reactions in which water is added to the organic molecule being broken down into simple structures.

The enzymes located inside the cell are known as intracellular enzymes. These enzymes include hydrolytic and oxidative enzymes. It is important to realize that all the oxidative reactions occur within the cell and not outside, for it is the oxidative reactions which bring about the energy changes required by the cell.

Hydrolytic Enzymes

Hydrolytic enzymes are concerned with the addition and the removal of water from the organic matter undergoing reaction. The most common hydrolytic reactions are concerned with the addition of water to complex organic molecules causing them to split into their basic components. This

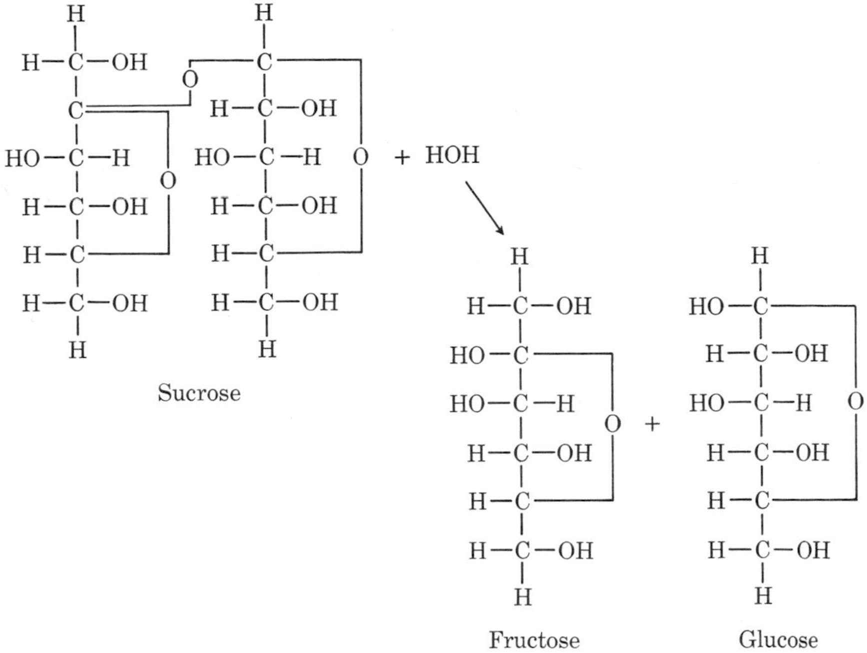

H—C—OH
C
HO—C—H
H—C—OH
H—C
H—C—OH
H

H—C—OH
HO—C—H
H—C—OH
H—C
H—C—OH
H

O + HOH

Sucrose

H—C—OH
HO—C
HO—C—H
H—C—OH
H—C
H—C—OH
H

+

HO—C
H—C—OH
HO—C—H
H—C—OH
H—C
H—C—OH
H

Fructose   Glucose

(*a*) Hydrolysis of sucrose

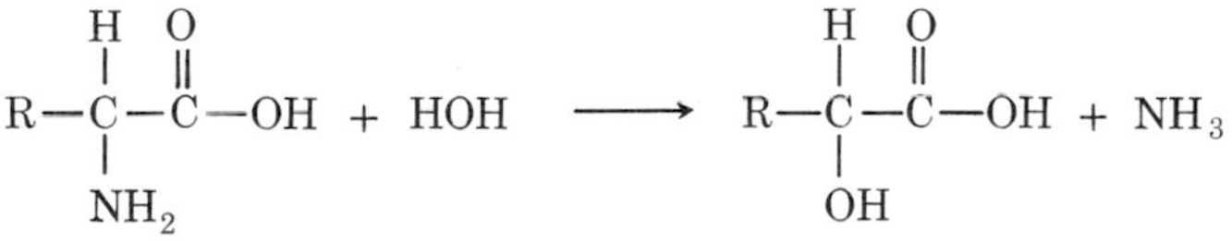

$$R—\underset{NH_2}{\overset{H}{\underset{|}{C}}}—\overset{O}{\overset{\|}{C}}—OH + HOH \longrightarrow R—\underset{OH}{\overset{H}{\underset{|}{C}}}—\overset{O}{\overset{\|}{C}}—OH + NH_3$$

(*b*) Hydrolysis of α-amino acids

FIG. 7-7. Typical hydrolytic enzyme reactions.

is the prime enzyme reaction occurring outside of the cell. Figure 7-7 shows examples of the hydrolytic split. Polysaccharides can be enzymatically hydrolyzed into the basic monosaccharides. Proteins can be hydrolyzed into amino acids. Fats are hydrolyzed to glycerol and fatty acids. The hydrolytic reaction is a low-energy reaction.

The second type of hydrolysis is equally as important as the first but

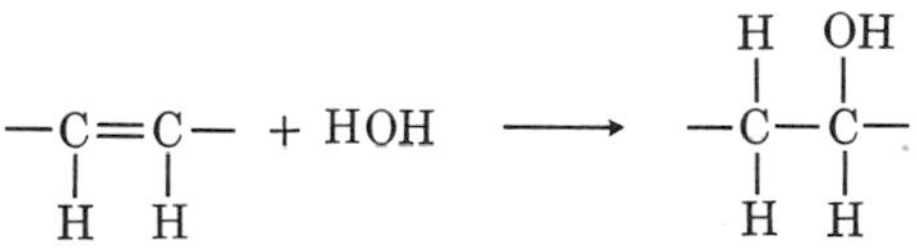

$$—\underset{H}{\overset{H}{C}}=\underset{H}{\overset{}{C}}— + HOH \longrightarrow —\underset{H}{\overset{H}{C}}—\underset{H}{\overset{OH}{C}}—$$

FIG. 7-8. Addition of water to ethylene group.

it is primarily an intracellular reaction since it is tied up with the energy systems within the cell. This reaction consists of the addition of water between two atoms aving a double bond between them. Figure 7-8 illustrates the mechanism of adding water to an ethylene group.

Oxidation-Reduction Enzymes

The oxidation-reduction enzymes are the intracellular enzymes which are responsible for the breakdown of substrate for energy and the synthesis of new protoplasm. The major oxidation-reduction enzymes are concerned with the transfer of hydrogen. Removal of hydrogen from

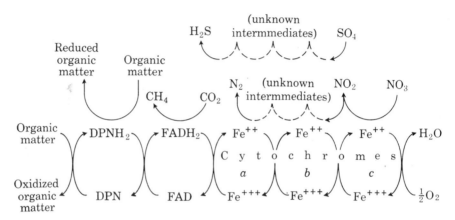

FIG. 7-9. Oxidation-reduction enzyme reactions.

organic matter results in its oxidation, while addition of hydrogen to organic matter results in its reduction. In aerobic biological systems oxygen is the ultimate hydrogen acceptor. In anaerobic biological systems the hydrogen acceptors include any oxidized organic matter, nitrates, nitrites, sulfates, and carbon dioxide (Fig. 7-9).

A second set of oxidation-reduction enzymes is concerned with the reactions of peroxides. Very little is known about the peroxide enzymes

$$H-O-O-H \xrightarrow{\text{Catalase}} H-O-H + \tfrac{1}{2}O_2$$

$$\text{Hydrogen peroxide} \qquad\qquad \text{Water} \quad \text{Oxygen}$$

FIG. 7-10. Peroxide reaction.

other than their ability to decompose peroxides into water and free oxygen (Fig. 7-10). The decomposition of hydrogen peroxide is brought by the enzyme, catalase. While much research has been done on catalase in bacteria, nothing is known of why it exists in bacteria. Recent research into the mechanism of photosynthesis indicates its potential use in algae and in autotrophic bacteria. This reaction will be discussed later.

Classification

Enzyme classification is confusing to say the least. There are several different schemes for naming enzymes in use today, all leading to confusion. The two predominant schemes are naming the enzyme (1) for the type reaction it produces and (2) for the material undergoing reaction. The enzyme nomenclature has a distinctive ending, -*ase*. Examples of type reactions are:

1. Hydrolase—hydrolytic enzymes
2. Oxidase—oxidative enzymes
3. Reductase—reduction enzymes

Examples of nomenclature based on the material undergoing reaction are:

1. Sucrase—hydrolysis of sucrose
2. Lipase—hydrolysis of fats
3. Amylase—hydrolysis of starch

A few enzymes have retained their original names. These enzymes include rennin, pepsin, and trypsin. All these enzymes are concerned with the hydrolysis of proteins.

It can easily be seen that one enzyme can have two names. There are times when both names are used but it leads to confusion. It is best to use the name of the enzyme which is the most inclusive. Thus, while the enzyme sucrase is also a hydrolase, the specific term sucrase should always be used to describe the hydrolysis of sucrose to glucose and fructose.

Source of Enzymes

The nucleus is the source of the enzymes within the cell. Certain enzymes are produced continuously by the nucleus, regardless of the biochemical environment in which the cell is growing. These enzymes are known as constitutive enzymes since they constitute the permanent portion of the cells' enzymes. The constitutive enzymes include all the enzymes concerned with synthesis of protoplasm and the normal energy yielding reactions. If the cell is placed in a nutrient medium not normal for its growth, it is often able to adapt its enzyme systems to the new nutrients. The new enzymes have been designated as adaptive enzymes.

The adaptive enzymes are not permanent enzymes like the constitutive enzymes but rather are temporary enzymes. Once the stimulus for the adaptive enzyme has been removed, the cell ceases to produce the adaptive enzyme. Thus we see that adaptive enzyme formation is merely the cell's effort to survive when its normal nutrient source is replaced by a foreign nutrient.

Much research has been carried out on adaptive enzyme formation but much is still to be learned. It is possible to derive certain conclusions about adaptive enzyme formation by considering how the nucleus forms enzymes.

Enzyme Formation

Enzymes occupy a certain space within the cell. For this reason there can be only a limited quantity of enzymes within the cell. The regenerative nature of enzyme reactions permits a small quantity of enzymes to bring about many reactions. The fact that the enzyme is divided into two major parts is of considerable importance to the cell. It is possible to combine the different coenzymes with the different apoenzymes to bring about an infinite number of reaction combinations.

The apoenzyme portion of the enzyme is the highly variable portion of the enzyme and is made up of various combinations of amino acids. The nucleus joins the amino acids to form the necessary apoenzymes and then combines the apoenzyme with desired coenzyme to form the desired enzyme. The need for many complex amino acid fragments to make the apoenzyme has led to the concept of the "protein pool" where amino acid fragments are stored until needed. When the enzyme is no longer needed, it is taken apart with the amino acid fragments being returned to the protein pool. While the protein pool does not exist as a specific structure, it represents a scheme by which enzyme synthesis can be made easily. In all probability the nucleus acts as the protein pool in which old enzymes are taken apart and new ones are created from the old parts.

Adaptive enzyme formation appears to be the temporary rearrangement of the apoenzymes to fit the unusual nutrient being metabolized. There is a limit to the microorganism's ability to produce adaptive enzymes. For the most part a microorganism can form adaptive enzymes for a few compounds related to the chemical structure of the compounds normally metabolized by the constitutive enzymes.

An apparent form of adaptive enzyme formation is use of enzymes in the synthesis system for energy metabolism. Normally, the synthesis enzymes are present in very low concentrations since many synthesis reactions are required for protoplasm. It is possible to add a nutrient normally used in the synthesis system or related to the synthesis system. Without its normal energy source the microorganism builds up those enzymes related to the synthesis system which are necessary to tie the new nutrient to the normal energy system. Once the new nutrient has been metabolized, the enzyme system will return to its normal concentration required for synthesis only. This latter form of enzyme adaptation is the most common type of adaptation in biological waste treatment systems.

CHAPTER 8

Metabolic Reactions

Microorganisms are actually small chemical factories in which raw materials are processed through a series of reactions to yield a finished product. The microbial end product is protoplasm. But the raw materials can be almost anything in nature.

The autotrophic bacteria utilize inorganic compounds entirely to produce an organic end product. Inorganic salts furnish the building blocks, as well as the energy. Algae also utilize inorganic materials for synthesizing protoplasm, but they use sunlight for energy.

Most microorganisms are heterotrophic, utilizing organic matter for both the building blocks and for energy. Microorganisms can start with one single raw material and produce the hundreds of different molecules making up protoplasm. When one considers the chemical transformations which take place within the microorganisms, one cannot help but marvel at the wonders of these little chemical factories.

It is important for the sanitary microbiologist to have a complete understanding of the metabolic transformations which occur within the microorganisms. Only by such a knowledge can the microbiologist control biological growth and metabolism. Such control is essential when dealing with the helpful saprophytic microbes in waste disposal systems or with the deadly pathogenic microbes.

Chemical Structure

The key to metabolic reactions lies in the chemical structure of the compounds being metabolized. This point has already been discussed with regard to enzymes but can stand emphasizing again.

All the biochemical reactions are enzyme reactions with these reactions a function of chemical structure. Chemical structure determines the solubility of the compound, the spatial configurations, the attraction or repulsion of molecules, and other properties which may or may not affect reaction.

Key Radicals

Biological reactions are orderly, following a fixed pattern. One of the most frustrating aspects of biochemistry is trying to find out this reaction pattern. Strangely enough, most microbiologists have been so intrigued by unusual reactions in unusual microorganisms that very little is known about the general metabolic patterns within the common microorganisms in nature. In spite of the lack of mass evidence of metabolic activity, it is possible to piece together a general metabolic picture that covers the common microbes.

The metabolic reactions begin at the beginning, but what part of a molecule is the beginning? The beginning is determined by the presence of certain key reactive chemical groups, radicals. The most reactive radical is the carboxyl group $-\overset{\overset{\displaystyle O}{\|}}{C}-OH$. In all organic acids of six carbons or less, the initial reaction will occur at the carboxyl group. Naturally, one cannot help but wonder what happens above six carbons if the carboxyl group is so reactive. Actually, the carboxyl group is so reactive that it has a greater affinity for water and permits the biological reaction to begin at another point. This is most obvious with long chain fatty acids and will be discussed in detail in a later section of this chapter.

Other chemical groups which undergo biological reactions in decreasing order of reaction are the aldol carbonyl $-\overset{\overset{\displaystyle O}{\|}}{C}H$, the keto carbonyl $-\overset{\overset{\displaystyle O}{\|}}{C}-$, the hydroxyl ($-OH$), the amino ($-NH_2$), the thiol ($-SH$), and the methyl ($-CH_3$). There are other radicals to be considered, the amide $-\overset{\overset{\displaystyle O}{\|}}{C}-NH_2$, the ester $-\overset{\overset{\displaystyle O}{\|}}{C}-O-$, the sulfate ($-O-SO_3H$), the sulfonate ($-SO_3H$), the chloro ($-Cl$), the ether ($-O-$), etc. These radicals are either related to the first group of radicals or their positions have not been fully determined.

Since each chemical group affects every other chemical group, it is not possible to state too many basic rules of metabolism. It is best to demonstrate the metabolic patterns by showing various metabolic reactions. The microbes carry out two reactions simultaneously: (1) degradation of matter for energy and (2) synthesis of protoplasm. These reactions are all interrelated but will be discussed in separate chapters. This chapter will be concerned with the general metabolic transformations. It should always be realized that the reactions shown are reversible. While most of the reactions will be concerned with degradation, the synthesis reaction is merely the reverse reaction.

Saturated Hydrocarbon

The low solubility of the saturated hydrocarbons makes them very difficult to metabolize but there are many bacteria which can degrade these molecules. The simplest molecule is methane. It is the hardest hydrocarbon to degrade and the only saturated hydrocarbon produced by bacteria. Methane is formed by the reduction of carbon dioxide.

$$CO_2 + 8H_2 \rightarrow CH_4 + 2H_2O \qquad (8\text{-}1)$$

The oversimplification of this reaction is astounding. The hydrogen is not molecular hydrogen as shown but rather is on an energy enzyme, probably diphosphopyridine nucleotide (DPN). A postulated reaction is shown in Fig. 8-1.

In this scheme carbon dioxide is reduced to formic acid which in turn is reduced to the hydrate of formaldehyde. Formaldehyde is reduced to methanol. The next step of the reaction has not been established yet, but based on normal reactions an unknown reducing compound reacts with the methanol, liberating water. The methyl complex formed is reduced to methane while the unknown compound is regenerated. The oxidation of methane probably follows the reverse reaction.

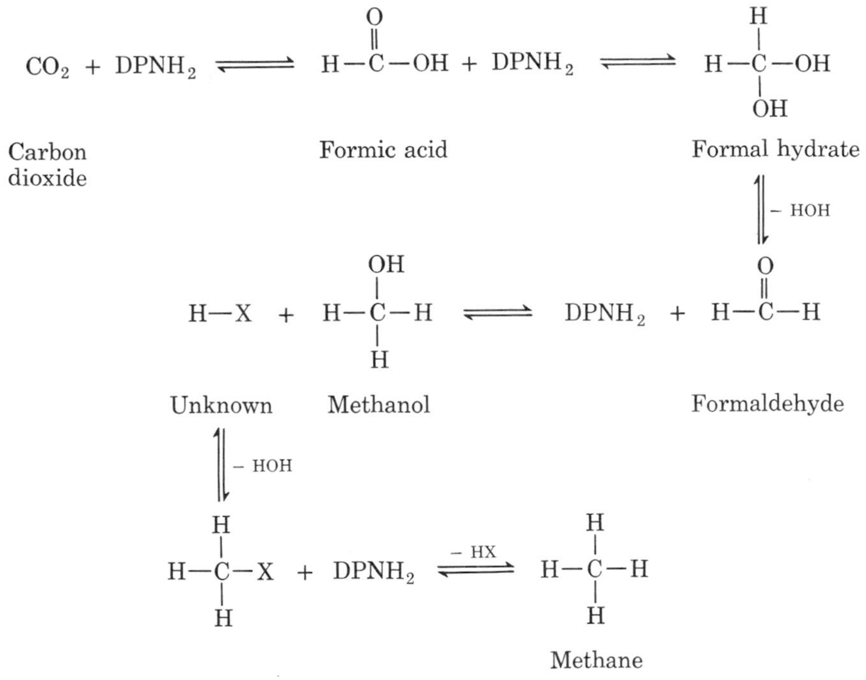

FIG. 8-1. Postulated mechanism of methane formation.

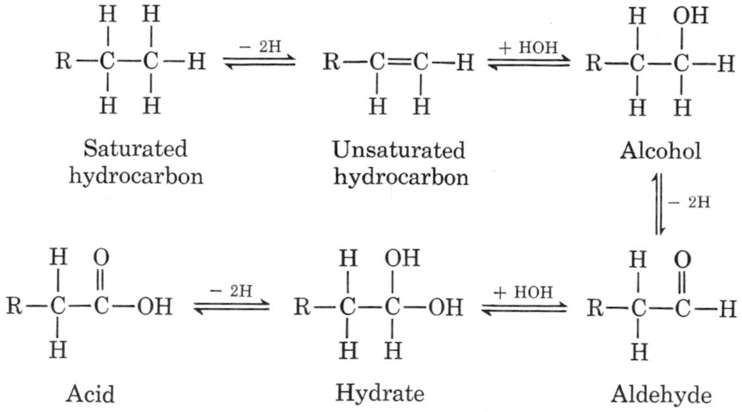

Fig. 8-2. Oxidation of saturated hydrocarbon.

In the metabolism of higher saturated hydrocarbons there is no need for a second compound to give up a hydrogen atom. Both hydrogen atoms are obtained from the same molecule on adjacent carbon atoms. The saturated hydrocarbon is converted to an unsaturated hydrocarbon which reacts with water to form an alcohol. The alcohol is oxidized to the aldehyde and on to the acid. This pattern is the basic metabolic pathway and is shown in Fig. 8-2.

If the starting material is ethane, acetate would be the acid produced. Acetate is the key intermediate in all biological metabolism and for this reason will be discussed in detail in the next section.

If the starting material contains three carbons or more, the acid formed must be further metabolized. The currently accepted scheme of metabolism is by beta oxidation in which the acids are split into two carbon fragments, acetate. The enzyme system involved in splitting the carbon chain contains CoA. CoA reacts with the acid at the sulfhydryl group to form the CoA-acid complex. Desaturation occurs between the alpha and beta carbons. Water is added to the molecule to form a β-hydroxy acid. Oxidation forms a keto acid which is split by CoA to form acetyl-CoA and an acid-CoA complex, now two carbon atoms shorter. The acetyl-CoA can be oxidized to carbon dioxide and water or synthesized into cellular components with liberation of CoA in either case for further reaction. The scheme of beta oxidation is shown in Fig. 8-3.

Acetate is always the end product of oxidation of organic acids containing an even number of carbon atoms. In organic acids containing an odd number of carbon atoms, acetate is split out until the final single-carbon fragment is split off, formyl-CoA $HC\overset{O}{\underset{||}{}}$—SCoA. The formyl-CoA

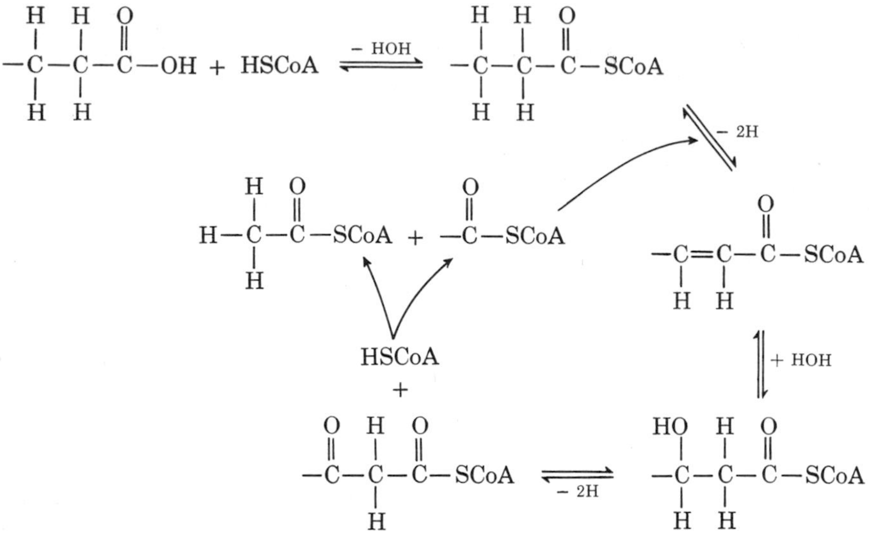

$$\begin{array}{c}
\text{H} \quad \text{H} \quad \text{O} \\
| \quad\; | \quad\; \| \\
-\text{C}-\text{C}-\text{C}-\text{OH} + \text{HSCoA} \;\xrightleftharpoons[\;]{-\,\text{HOH}}\; -\text{C}-\text{C}-\text{C}-\text{SCoA} \\
| \quad\; | \qquad\qquad\qquad\qquad\qquad\qquad\quad | \quad\; | \\
\text{H} \quad \text{H} \qquad\qquad\qquad\qquad\qquad\qquad\;\; \text{H} \quad \text{H}
\end{array}$$

$$\text{H}-\overset{\text{H}}{\underset{\text{H}}{\text{C}}}-\overset{\text{O}}{\text{C}}-\text{SCoA} \;+\; -\overset{\text{O}}{\text{C}}-\text{SCoA}$$

$$-\text{C}=\text{C}-\overset{\text{O}}{\text{C}}-\text{SCoA}$$ (with -2H and $+\text{HOH}$)

$$\text{HSCoA} +$$

$$-\overset{\text{O}}{\text{C}}-\overset{\text{H}}{\underset{\text{H}}{\text{C}}}-\overset{\text{O}}{\text{C}}-\text{SCoA} \;\xrightleftharpoons[-2\text{H}]{}\; -\overset{\text{HO}}{\underset{\text{H}}{\text{C}}}-\overset{\text{H}}{\underset{\text{H}}{\text{C}}}-\overset{\text{O}}{\text{C}}-\text{SCoA}$$

Fig. 8-3. Beta oxidation cycle for fatty acid metabolism.

is readily hydrolyzed to formic acid and oxidized to carbon dioxide and water.

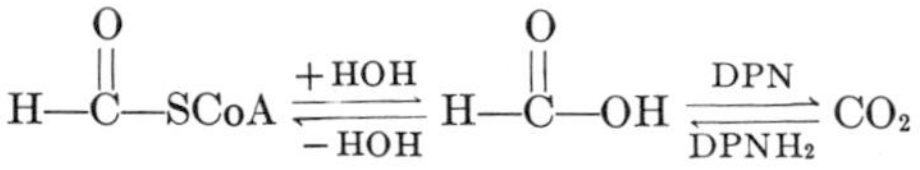

$$\text{H}-\overset{\text{O}}{\text{C}}-\text{SCoA} \;\xrightleftharpoons[-\,\text{HOH}]{+\,\text{HOH}}\; \text{H}-\overset{\text{O}}{\text{C}}-\text{OH} \;\xrightleftharpoons[\text{DPNH}_2]{\text{DPN}}\; CO_2$$

The importance of this reaction cannot be overemphasized since it explains how carbon dioxide can be introduced into metabolic reactions. The latter half of this reaction is the same shown for methane metabolism in Fig. 8-1. While methane formation is highly specific, the introduction of carbon dioxide is a general reaction common to all microbes.

In presenting a pattern for metabolism of saturated hydrocarbons, the pattern has also been presented for the normal alcohols, the aldehydes, and the acids. It will be shown that almost all compounds follow this metabolic pattern or slight modifications of it. For this reason the sanitary microbiologist must know this metabolic pathway backwards and forwards.

Acetate

Acetate metabolism is very common with microorganisms and is the key intermediate for energy and synthesis. Large molecules are built up from condensation of acetate and energy is obtained by its oxidation to carbon dioxide.

Current concepts of acetate oxidation indicate that acetate can condense with carbon dioxide through formyl-CoA to form pyruvic acid

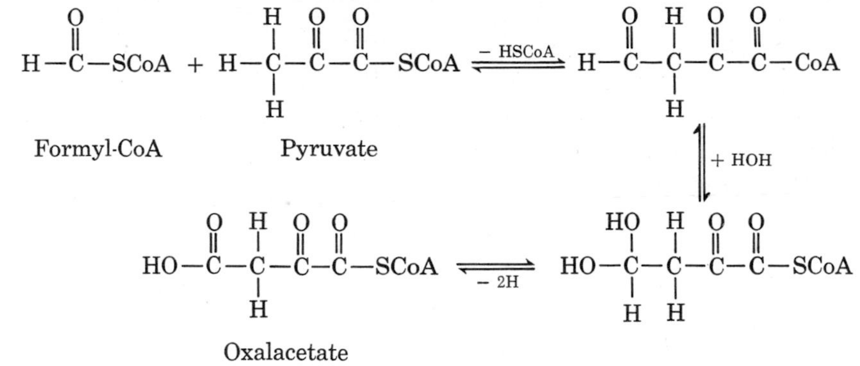

$$\underset{\text{Formyl-CoA}}{H-\overset{\displaystyle O}{\overset{\|}{C}}-SCoA} + \underset{\text{Pyruvate}}{H-\overset{\displaystyle H}{\underset{\displaystyle H}{\overset{|}{C}}}-\overset{\displaystyle O}{\overset{\|}{C}}-\overset{\displaystyle O}{\overset{\|}{C}}-SCoA} \xrightarrow{-\ HSCoA} H-\overset{\displaystyle O}{\overset{\|}{C}}-\overset{\displaystyle H}{\underset{\displaystyle H}{\overset{|}{C}}}-\overset{\displaystyle O}{\overset{\|}{C}}-\overset{\displaystyle O}{\overset{\|}{C}}-CoA$$

$\Updownarrow + HOH$

$$\underset{\text{Oxalacetate}}{HO-\overset{\displaystyle O}{\overset{\|}{C}}-\overset{\displaystyle H}{\underset{\displaystyle H}{\overset{|}{C}}}-\overset{\displaystyle O}{\overset{\|}{C}}-\overset{\displaystyle O}{\overset{\|}{C}}-SCoA} \underset{-\ 2H}{\rightleftharpoons} HO-\overset{\displaystyle HO}{\underset{\displaystyle H}{\overset{|}{C}}}-\overset{\displaystyle H}{\underset{\displaystyle H}{\overset{|}{C}}}-\overset{\displaystyle O}{\overset{\|}{C}}-\overset{\displaystyle O}{\overset{\|}{C}}-SCoA$$

FIG. 8-4. Condensation of carbon dioxide with pyruvic acid.

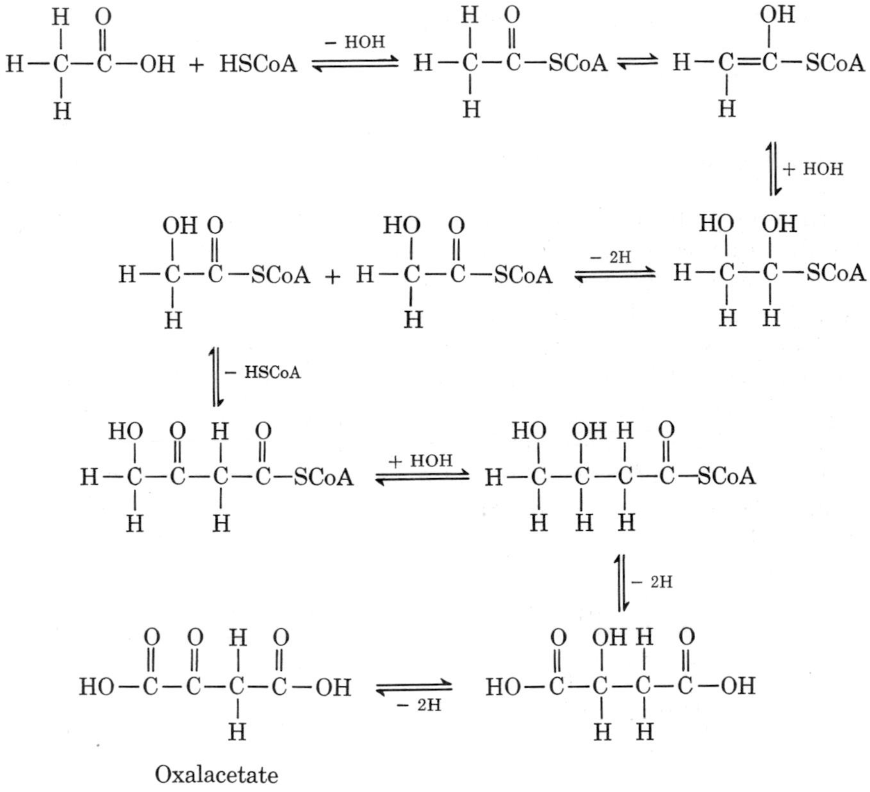

$$H-\overset{\displaystyle H}{\underset{\displaystyle H}{\overset{|}{C}}}-\overset{\displaystyle O}{\overset{\|}{C}}-OH + HSCoA \xrightarrow{-\ HOH} H-\overset{\displaystyle H}{\underset{\displaystyle H}{\overset{|}{C}}}-\overset{\displaystyle O}{\overset{\|}{C}}-SCoA \rightleftharpoons H-\overset{\displaystyle OH}{\overset{|}{C}}=\overset{}{\underset{\displaystyle H}{C}}-SCoA$$

$\Updownarrow + HOH$

$$H-\overset{\displaystyle OH}{\underset{\displaystyle H}{\overset{|}{C}}}-\overset{\displaystyle O}{\overset{\|}{C}}-SCoA + H-\overset{\displaystyle HO}{\underset{\displaystyle H}{\overset{|}{C}}}-\overset{\displaystyle O}{\overset{\|}{C}}-SCoA \xrightarrow{-\ 2H} H-\overset{\displaystyle HO}{\underset{\displaystyle H}{\overset{|}{C}}}-\overset{\displaystyle OH}{\underset{\displaystyle H}{\overset{|}{C}}}-SCoA$$

$\Updownarrow - HSCoA$

$$H-\overset{\displaystyle HO}{\underset{\displaystyle H}{\overset{|}{C}}}-\overset{\displaystyle O}{\overset{\|}{C}}-\overset{\displaystyle H}{\underset{\displaystyle H}{\overset{|}{C}}}-\overset{\displaystyle O}{\overset{\|}{C}}-SCoA \xrightarrow{+\ HOH} H-\overset{\displaystyle HO}{\underset{\displaystyle H}{\overset{|}{C}}}-\overset{\displaystyle OH}{\underset{\displaystyle H}{\overset{|}{C}}}-\overset{\displaystyle H}{\underset{\displaystyle H}{\overset{|}{C}}}-\overset{\displaystyle O}{\overset{\|}{C}}-SCoA$$

$\Updownarrow - 2H$

$$\underset{\text{Oxalacetate}}{HO-\overset{\displaystyle O}{\overset{\|}{C}}-\overset{\displaystyle O}{\overset{\|}{C}}-\overset{\displaystyle H}{\underset{\displaystyle H}{\overset{|}{C}}}-\overset{\displaystyle O}{\overset{\|}{C}}-OH} \underset{-\ 2H}{\rightleftharpoons} HO-\overset{\displaystyle O}{\overset{\|}{C}}-\overset{\displaystyle OH}{\underset{\displaystyle H}{\overset{|}{C}}}-\overset{\displaystyle H}{\underset{\displaystyle H}{\overset{|}{C}}}-\overset{\displaystyle O}{\overset{\|}{C}}-OH$$

FIG. 8-5. Postulated metabolism of acetic acid.

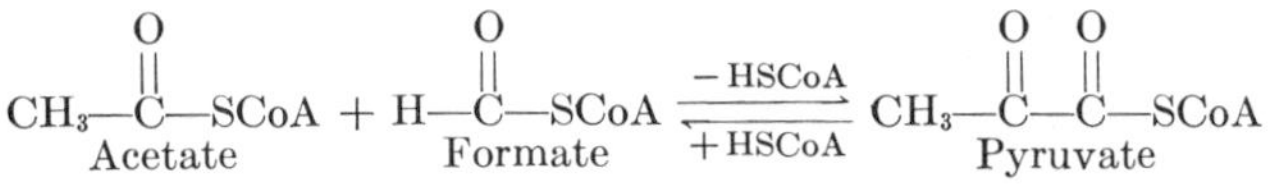

$$\underset{\text{Acetate}}{CH_3-\overset{O}{\overset{\|}{C}}-SCoA} + \underset{\text{Formate}}{H-\overset{O}{\overset{\|}{C}}-SCoA} \underset{+\,HSCoA}{\overset{-\,HSCoA}{\rightleftharpoons}} \underset{\text{Pyruvate}}{CH_3-\overset{O}{\overset{\|}{C}}-\overset{O}{\overset{\|}{C}}-SCoA}$$

As always the acids exist as the CoA complex since this is the reactive form. Pyruvate condenses with carbon dioxide and forms oxalacetate on oxidation (Fig. 8-4).

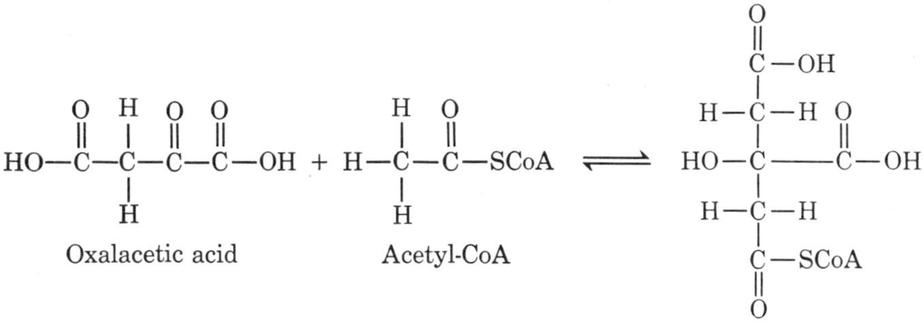

FIG. 8-6. Condensation of acetyl-CoA with oxalacetic acid to form citric acid.

The only problem with this reaction is that it requires an energy source for the reaction to occur. Unfortunately, in a pure solution of acetate, there is no available energy for the carbon dioxide condensation and it is highly unlikely that the reaction will occur.

Energy must be obtained from the direct oxidation of acetate. Unfortunately, very little work has been done on this phase of metabolism and little is known about it. The reactions in Fig. 8-5 are postulated as a more probable mechanism for acetate metabolism.

Acetate, as the acetyl-CoA, picks up water to form a dihydroxy complex which loses two hydrogen to form a glycolic complex. The glycolic-CoA

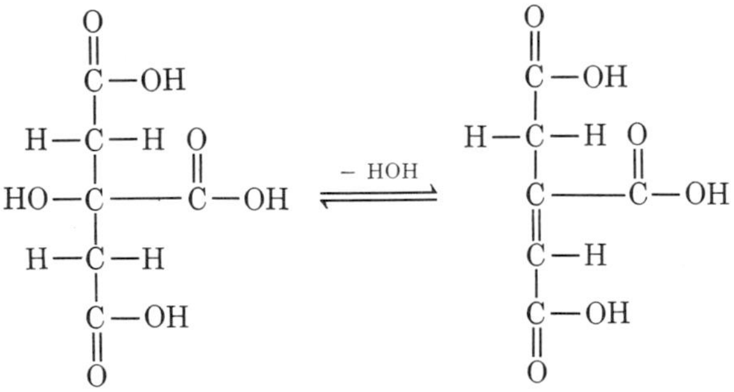

FIG. 8-7. Conversion of citric acid to cis-aconitic acid.

condenses with acetyl-CoA to form a four-carbon compound which is oxidized to malate and then to oxalacetate. This scheme allows the acetate to yield energy while being condensed to oxalacetate. Acetyl-CoA can condense with oxalacetate to form citrate (Fig. 8-6).

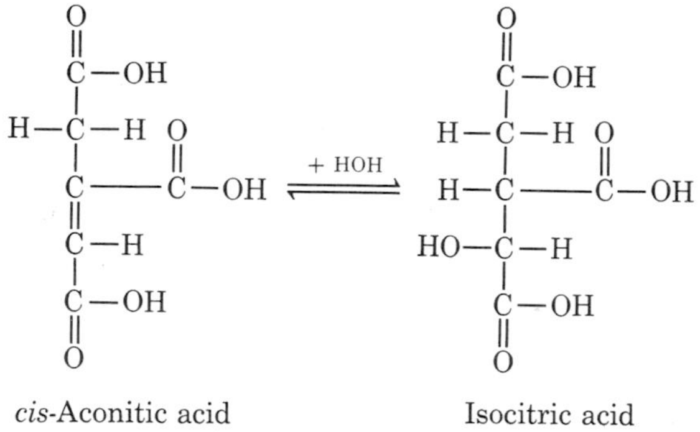

cis-Aconitic acid   Isocitric acid

FIG. 8-8. Hydrolytic conversion of cis-aconitic acid to isocitric acid.

Citrate loses water to form cis-aconitate (Fig. 8-7). The cis-aconitate takes up water but in a different position to form isocitrate (Fig. 8-8). Isocitrate is oxidized to oxalosuccinate (Fig. 8-9). Decarboxylation splits out carbon dioxide to form α-ketoglutarate (Fig. 8-10).

While the metabolic reactions to this point have been fairly well accepted, they follow such devious chemical reactions that one cannot but question their validity. It seems very strange that a 2C molecule must

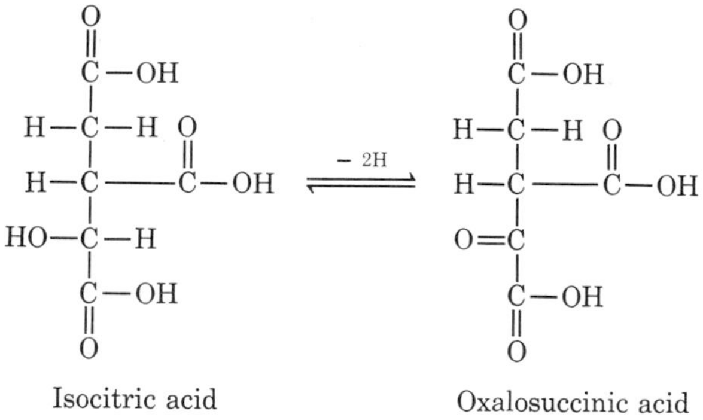

Isocitric acid   Oxalosuccinic acid

FIG. 8-9. Oxidation of isocitric acid to oxalosuccinic acid.

condense to form a more complex 6C molecule to be oxidized, too strange to be logical. But this does not answer how metabolism takes place. The simplest method would be to continue direct oxidation of glycolate to glyoxylate and to formate (Fig. 8-11). Unfortunately, no evidence exists for the existence of this scheme while some evidence exists against it. Further research is definitely needed on this point.

There is considerable evidence to point up the existence of condensation of acetate to form oxalacetate, citrate, *cis*-aconitate, isocitrate, ox-

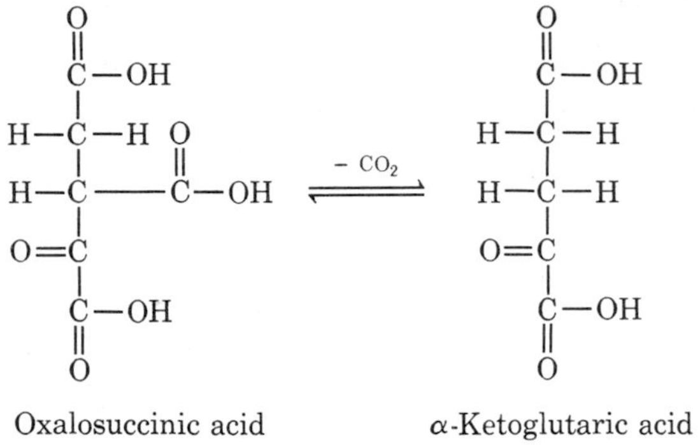

Oxalosuccinic acid   α-Ketoglutaric acid

FIG. 8-10. Decarboxylation of oxalosuccinic acid to form α-ketoglutaric acid.

alosuccinate, and α-ketoglutaric acid. But this condensation appears to be primarily related to synthesis metabolism and not to energy. These two schemes are distinctly separated and are yet related. Even in synthesis the reactions do not look entirely logical. The things that appear certain are the presence of succinate, glutarate, fumarate, malate, and oxalacetate. The condensation as shown previously in Fig. 8-6 would

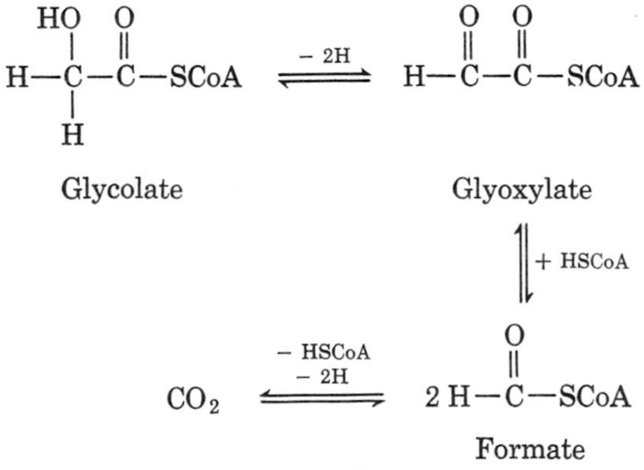

Glycolate   Glyoxylate

Formate

FIG. 8-11. Possible oxidation of acetate through glycolate.

yield oxalacetate which would be reduced to malate, dehydrated to fumarate, reduced to succinate, and condensed with carbon dioxide to α-ketoglutaric acid (Fig. 8-12). In this way it would be possible to eliminate the citrate to oxalosuccinate phase which has caused so much trouble to the microbiologist. Yet, until more evidence is available to the contrary, the citric acid cycle (Kreb cycle) will be accepted as the terminal oxidation scheme for acetate. The complete citric acid cycle is given in Fig. 8-13.

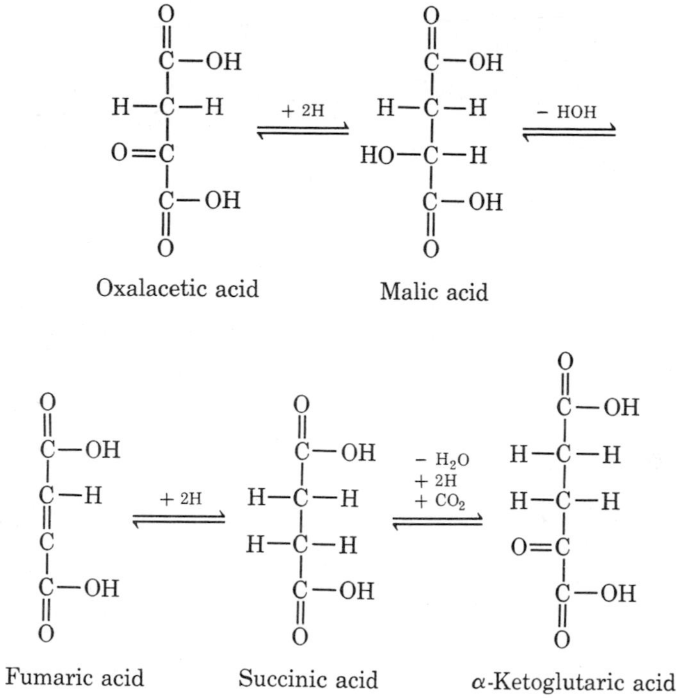

Oxalacetic acid   Malic acid

Fumaric acid   Succinic acid   α-Ketoglutaric acid

Fig. 8-12. Formation of α-ketoglutaric acid from oxalacetic acid.

Alcohols

The metabolic reactions of the normal alcohols have already been given, but there is the problem of secondary and tertiary alcohols. The secondary alcohols R—$\overset{\text{H}}{\underset{\text{OH}}{\text{C}}}$—R are oxidized to the ketone which is hydrated to the dihydroxy alcohol (Fig. 8-14).

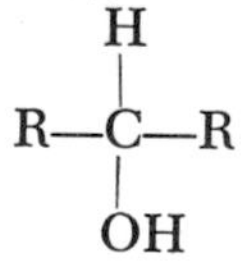

The tertiary alcohol poses a definite problem. This is most obvious in *t*-butanol,

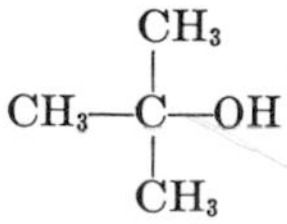

$$CH_3-\overset{\text{CH}_3}{\underset{\text{CH}_3}{\text{C}}}-OH$$

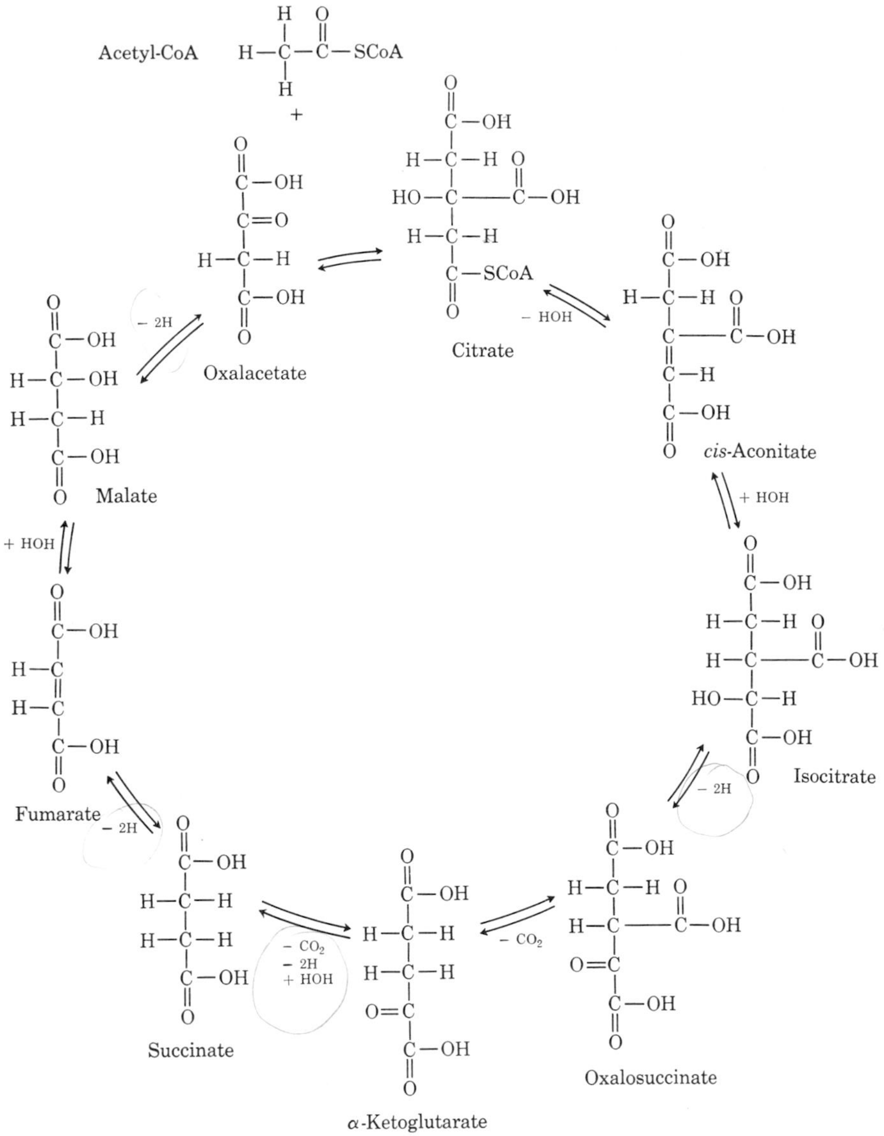

Acetyl-CoA

Oxalacetate

Citrate

Malate

cis-Aconitate

Isocitrate

Fumarate

Succinate

α-Ketoglutarate

Oxalosuccinate

$-2H$

$+HOH$

$-HOH$

$+HOH$

$-2H$

$-2H$

$-CO_2$
$-2H$
$+HOH$

$-CO_2$

FIG. 8-13. Citric acid cycle.

There is no way for hydrogen to be split out of the molecule except in adjacent molecules such as methane oxidation. This mechanism of metabolism does not occur normally except in highly specialized microorganisms capable of degrading t-butanol. This means that dehydration followed by hydration with the shift in the hydroxyl group such as pro-

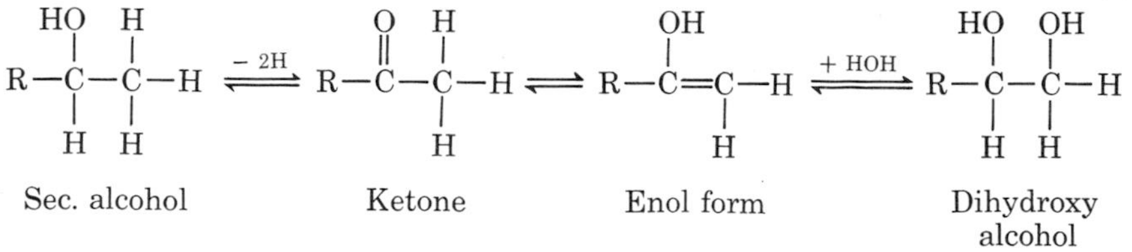

$$
\underset{\substack{\text{Sec. alcohol}}}{R-\overset{\displaystyle HO}{\underset{\displaystyle H}{\overset{|}{\underset{|}{C}}}}-\overset{\displaystyle H}{\underset{\displaystyle H}{\overset{|}{\underset{|}{C}}}}-H}
\;\underset{\text{}}{\overset{-2H}{\rightleftharpoons}}\;
\underset{\substack{\text{Ketone}}}{R-\overset{\displaystyle O}{\overset{\|}{C}}-\overset{\displaystyle H}{\underset{\displaystyle H}{\overset{|}{\underset{|}{C}}}}-H}
\;\rightleftharpoons\;
\underset{\substack{\text{Enol form}}}{R-\overset{\displaystyle OH}{\overset{|}{C}}=\overset{\displaystyle H}{\overset{|}{C}}-H}
\;\overset{+\,HOH}{\rightleftharpoons}\;
\underset{\substack{\text{Dihydroxy}\\\text{alcohol}}}{R-\overset{\displaystyle HO}{\underset{\displaystyle H}{\overset{|}{\underset{|}{C}}}}-\overset{\displaystyle OH}{\underset{\displaystyle H}{\overset{|}{\underset{|}{C}}}}-H}
$$

FIG. 8-14. Metabolism of secondary alcohol to a dihydroxy alcohol.

posed in converting citrate to *cis*-aconitate to isocitrate does not occur readily.

The dihydroxy alcohols, the glycols, are easily oxidized to the corresponding α-hydroxy acid which is converted to the keto acid. The keto acid is split to formyl-CoA and the acid-CoA, as shown in Fig. 8-15.

The trihydroxy alcohol, glycerol, is metabolized by a different scheme than for the dihydroxy alcohols. Glycerol reacts with ATP to form glycerol-PO_4 which is dehydrogenated to dihydroxyacetone-PO_4. Dihydroxyacetone-PO_4 is in equilibrium with 3-PO_4 glyceraldehyde which reacts with inorganic phosphate and DPN to form 1,3-PO_4 glyceric acid. ADP takes phosphate from 1,3-PO_4 glyceric acid to form 3-PO_4 glyceric acid. The PO_4 group migrates to carbon 2 to form 2-PO_4 glyceric acid. Water is split out to form 2-PO_4 enolpyruvic acid. ADP takes phosphate to form

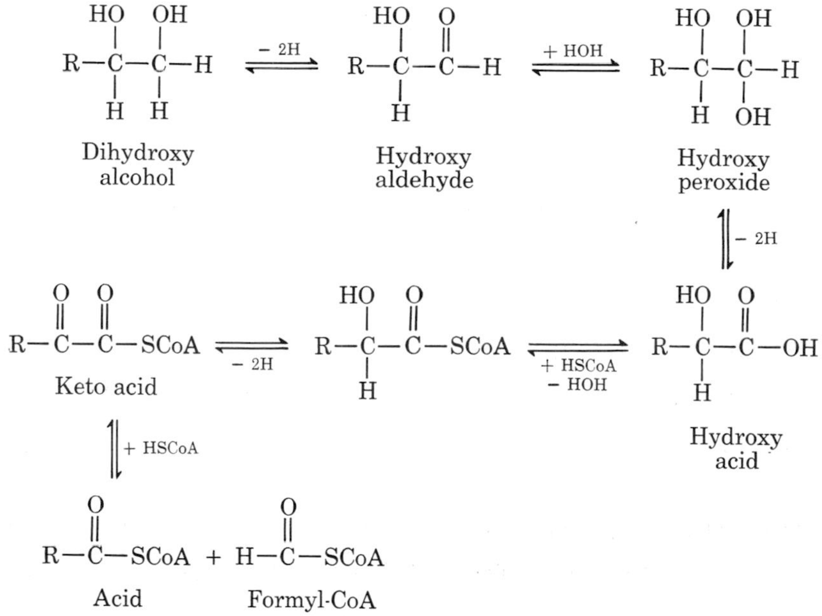

FIG. 8-15. Metabolism of dihydroxy alcohols.

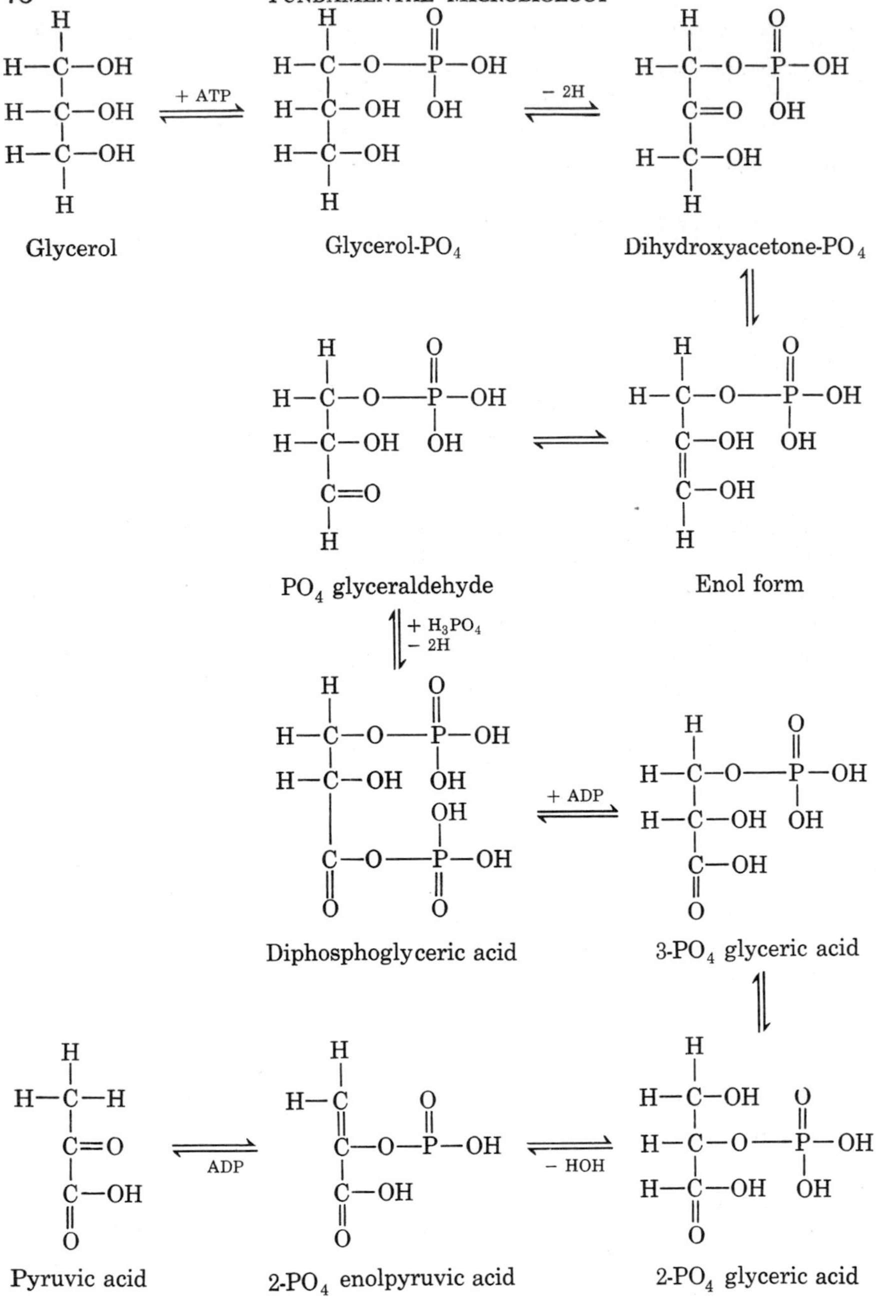

FIG. 8-16. Metabolism of glycerol to pyruvic acid.

enolpyruvate which is in equilibrium with pyruvate. These reactions are shown in Fig. 8-16.

The above series of reactions have been quite well elucidated, but one cannot help but wonder why the microorganism follows such a complex series of reactions. It would seem that a simpler scheme would be the oxidation of glycerol to glyceraldehyde and then to glyceric acid. Dehydration of glyceric acid would yield enolpyruvate directly (Fig. 8-17).

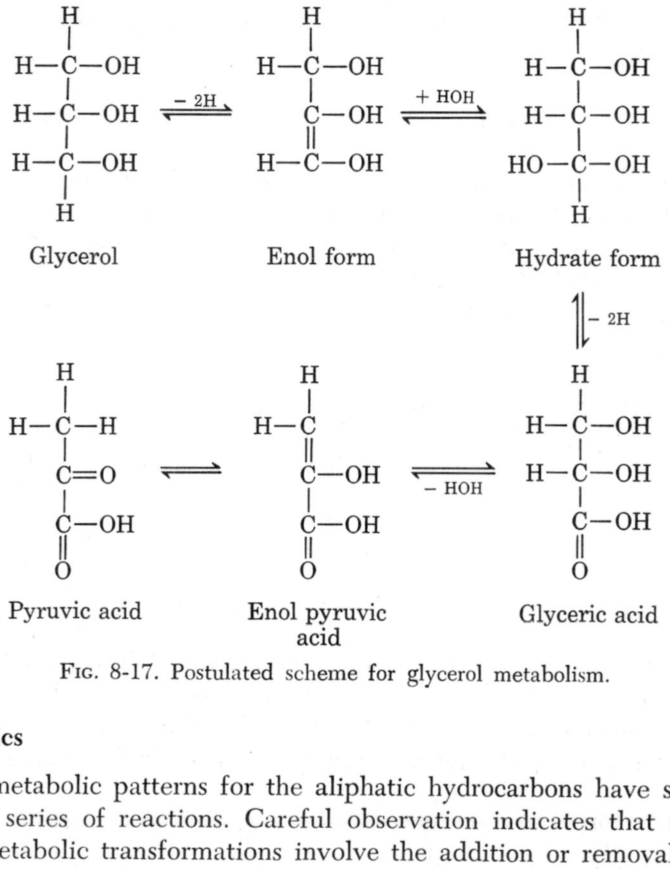

FIG. 8-17. Postulated scheme for glycerol metabolism.

Aromatics

The metabolic patterns for the aliphatic hydrocarbons have shown a definite series of reactions. Careful observation indicates that most of these metabolic transformations involve the addition or removal of hydrogen, water, and carbon. These basic reactions hold for all metabolic patterns and should be thoroughly understood.

The aromatic compounds undergo generally the same type of reactions but in a slightly different pattern. The reason for this difference lies in the chemistry of the benzene ring. The benzene ring is a six-carbon unsaturated compound of very unusual stability. The stability of the benzene ring is due to the resonance between two basic forms (Fig. 8-18).

In spite of its stability, benzene can be degraded. Needless to say with three unsaturated points, the first step would be hydration followed by dehydrogenation to yield phenol (Fig. 8-19).

The presence of a radical such as the hydroxyl group on the benzene ring changes the resonance within the benzene ring (Fig. 8-20). This

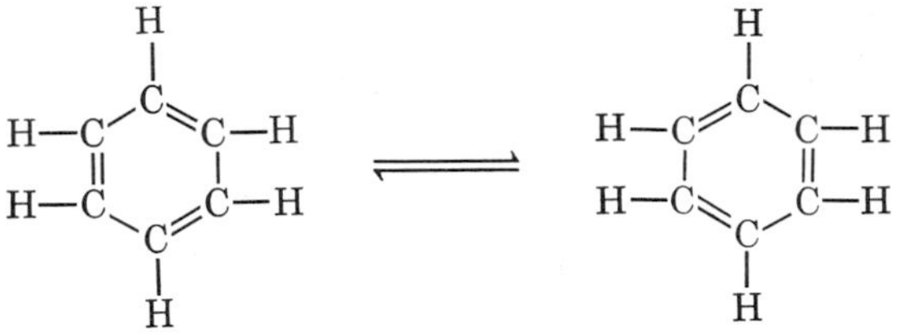

FIG. 8-18. Resonance forms of benzene.

change in resonance makes the ring more susceptible to further reaction. The next step in the reaction has not been fully elucidated, but the end product is known. Adding water and removing hydrogen from phenol produces catechol (Fig. 8-21).

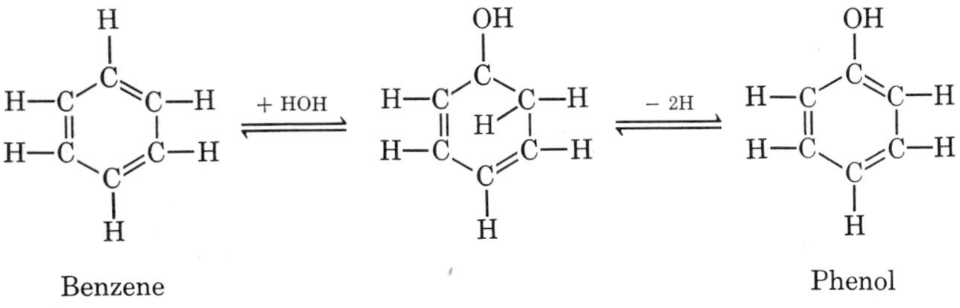

FIG. 8-19. Oxidation of benzene to phenol.

Catechol also has its own resonance forms (Fig. 8-22). The latter form is probably the reactive form. With the dicarbonyl radical in aliphatic hydrocarbons the carbon-carbon bond is split with CoA. The same type of reaction occurs with catechol (Fig. 8-23). The aldol carbonyl is easily

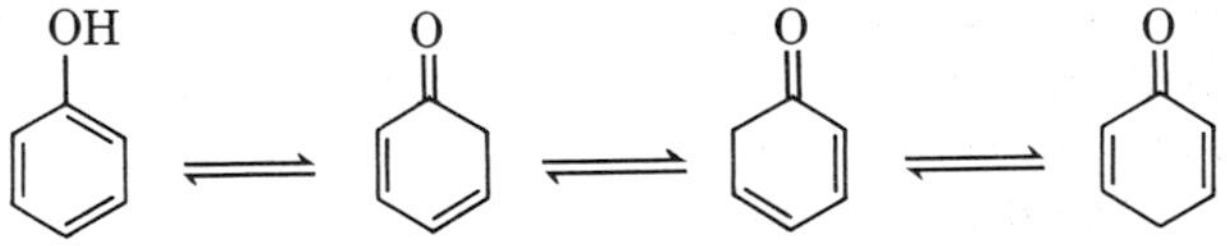

FIG. 8-20. Resonance forms of phenol.

oxidized to a carboxyl group to give 3,4-unsaturated adipic acid. Hydration of this molecule gives 3-hydroxyadipic acid which is dehydrated to 3-ketoadipic acid. Beta oxidation with CoA permits the formation of succinic acid and acetic acid (Fig. 8-24).

Succinate is supposed to be oxidized by the citric acid cycle, as shown in Fig. 8-13. But the problem remains as to what happens beyond oxalacetate. There is a need for acetate to come into the picture. One might point out that acetate is present and is what causes the oxalacetate to be converted to citrate and keeps the cycle going. One can only be

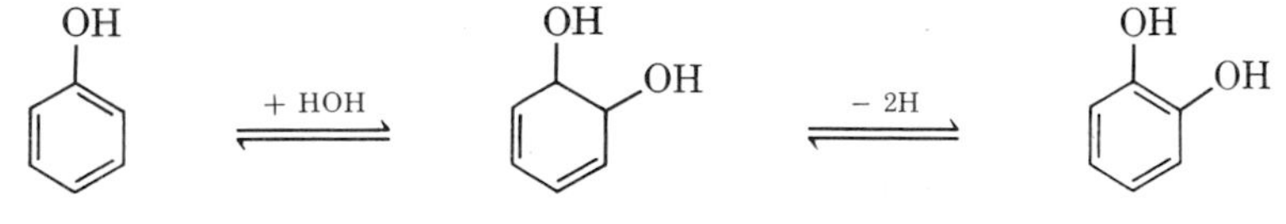

Fig. 8-21. Oxidation of phenol to catechol.

amused at such illogical logic. Careful examination of the reaction would easily indicate that the acetate would be oxidized to carbon dioxide but that the succinate would never get beyond oxalacetate if the citric acid cycle were the only pattern of oxidation. Obviously, there is a separate

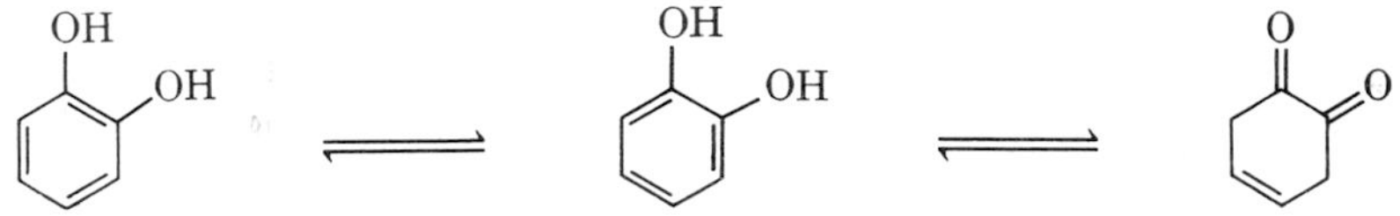

Fig. 8-22. Resonance forms of catechol.

terminal oxidation pattern for the dicarboxylic acids. Since most microbiologists have never been disturbed by the logic of how succinic acid was oxidized to carbon dioxide, nothing is known of this pattern of metabolism. It is known though that carbon dioxide can add to pyruvate

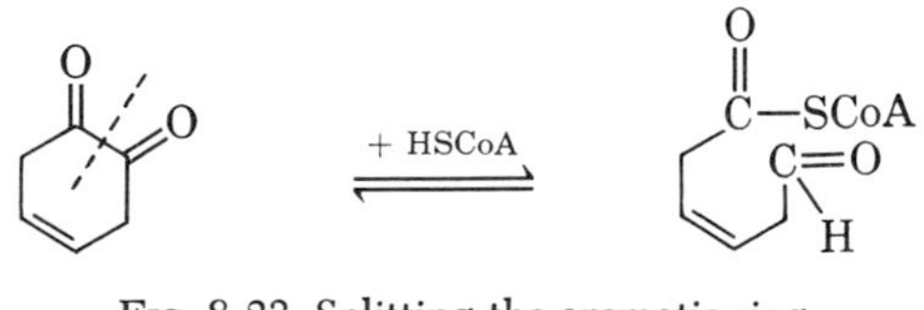

Fig. 8-23. Splitting the aromatic ring.

to yield oxalacetate. The reverse reaction is possible, as shown in Fig. 8-4.

The only problem with this reaction is that hydrogen is required for the split. Normally, the addition of hydrogen is an energy requiring reaction. For this reason it is doubtful if this reaction is the terminal oxidation pattern. There is a definite similarity between succinate and acetate metabolism, but no one has ever been able to put the two together.

One thing for certain, succinate is not formed by a back-to-back condensation of acetate. This is definitely an area for further research.

Carbohydrates

Carbohydrates are one of the most readily available metabolic materials. All microorganisms have the ability to degrade and to synthesize carbohydrates, even though some microorganisms cannot metabolize ex-

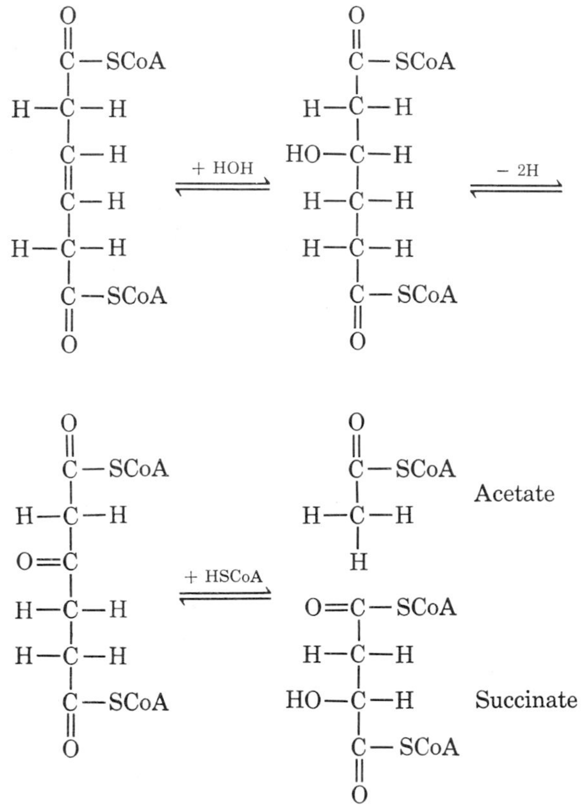

FIG. 8-24. Oxidation of 3,4-unsaturated adipic acid.

ternal sources of carbohydrates. Since carbohydrates are used so extensively for energy, the metabolic reactions will be given from the standpoint of degradation.

The two basic carbohydrate molecules are pentoses and hexoses. The pentoses are five-carbon sugars, $C_5H_{10}O_5$, while the hexoses are six-carbon sugars, $C_6H_{12}O_6$. The problem with carbohydrates lies in their molecular configuration. They have a definite ring structure but a resonance form to yield an aldol carbonyl or a keto carbonyl (Fig. 8-25).

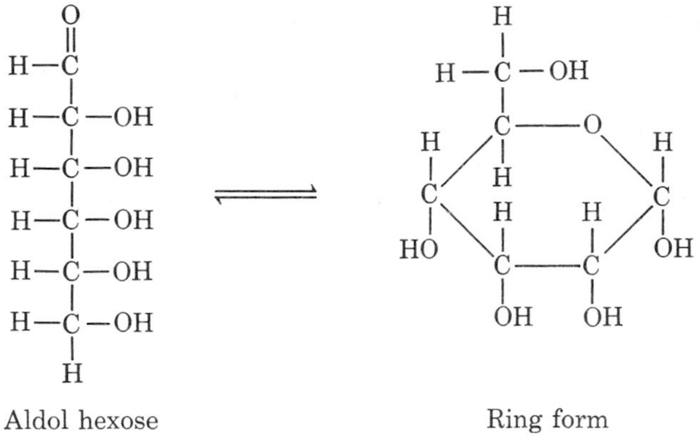

Aldol hexose   Ring form

FIG. 8-25. Structure of aldol hexose.

Thus, the carbohydrates have the properties of a ring and a straight chain. The keto hexose is shown in Fig. 8-26. It is important to note that both the aldol hexose and keto hexose have the same enol form.

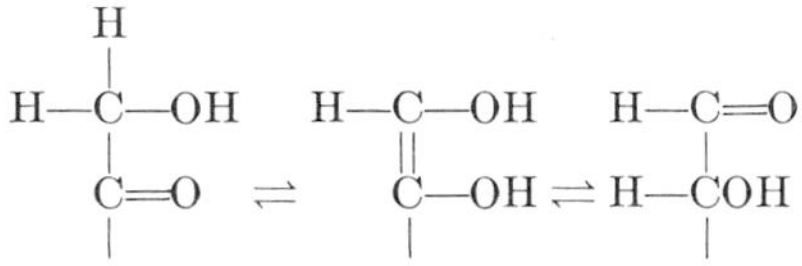

The difference between hexose molecules lies in the position of the hydroxyl groups along the chain. The structural formulas of the four most common hexoses are given in Fig. 8-27.

One of the properties of carbohydrates is the ease with which they polymerize to form larger molecules. One molecule of glucose can join another glucose molecule to form maltose or cellibiose. When two carbo-

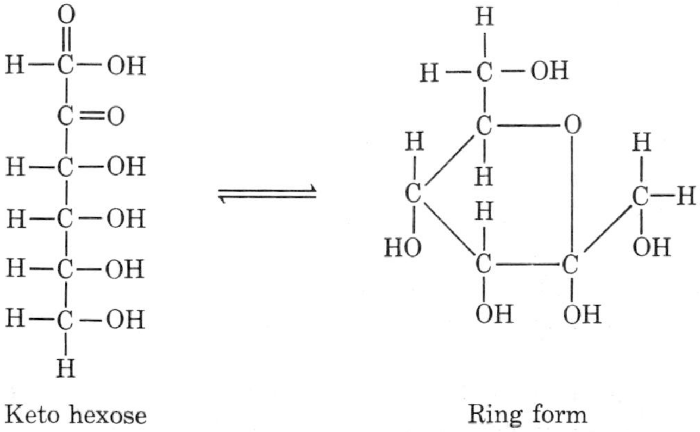

Keto hexose   Ring form

FIG. 8-26. Structure of keto hexose.

hydrate molecules condense, water is split out. The difference between maltose and cellibiose lies in the configuration of the two glucose molecules. In maltose the molecules join directly in an alpha linkage, but the molecules twist in cellibiose to form a beta linkage (Fig. 8-28). The minor modification in molecular structure has an important effect on the microorganisms' metabolic reactions, since some bacteria can metabolize maltose but not cellibiose.

Carbohydrates made of two monosaccharide molecules are called disaccharides. Important disaccharides other than maltose or cellibiose include sucrose (glucose + fructose) and lactose (glucose + galactose). Condensation of many carbohydrate molecules leads to polysaccharides such as glycogen, starch, and cellulose.

Metabolism of polysaccharides begins with hydrolysis of the complex

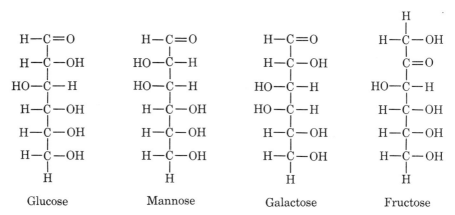

FIG. 8-27. Structure of four common hexoses.

polysaccharide to the monosaccharide. The polysaccharides are primarily glucose polymers so that carbohydrate metabolism is best demonstrated with glucose. The initial reaction is a phosphorylation in which ATP is reduced to ADP. The phosphorylation occurs on carbon 6. With carbon 6 tied up the reaction shifts to the upper end of the molecule (Fig. 8-29).

Glucose-6-PO_4 undergoes a dual pattern of metabolism, glycolysis and the pentose shunt. Glycolysis was the first pattern of metabolism elucidated for carbohydrates and will be presented first. As already indicated, the aldol and the keto sugars have the same enol form. It is not surprising then to see glucose-6-PO_4 converted to fructose-6-PO_4 (Fig. 8-30). The fructose-6-PO_4 picks up phosphate from ATP to form the 1,6-diphosphate (Fig. 8-31).

With both ends of the molecule tied up with phosphate, reaction shifts to the center of the molecule. The fructose 1,6-PO_4 is split into dihydroxyacetone-PO_4 and glyceraldehyde-PO_4 (Fig. 8-32).

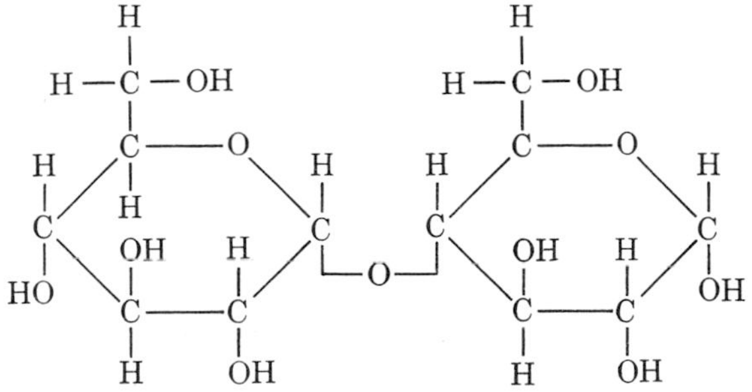

Maltose

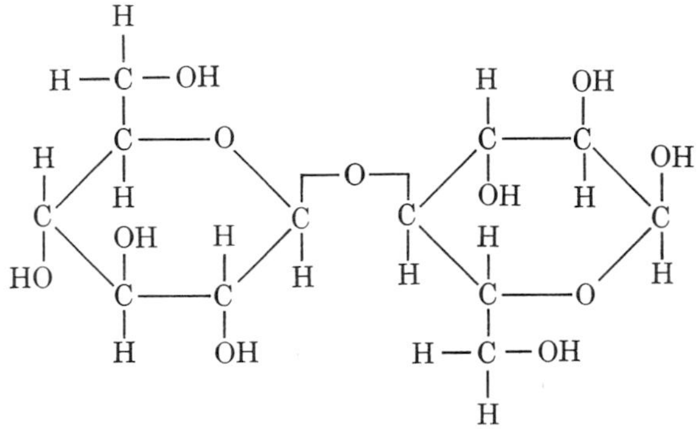

Cellibiose

Fig. 8-28. Structure of two disaccharides.

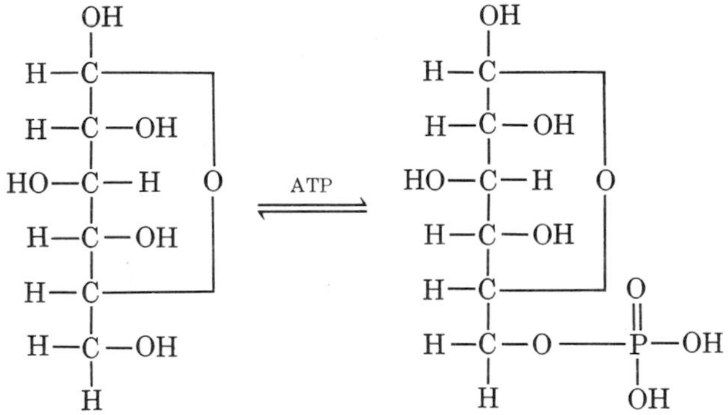

Glucose   Glucose-6·PO_4

Fig. 8-29. Phosphorylation of glucose.

The scheme of metabolism of dihydroxyacetone-PO$_4$ and glyceraldehyde has already been shown for that of glycerol.

The transformations involved in the splitting of the ring into two three-carbon fragments are fantastic. Although there is no doubt that a split

Glucose-6-PO$_4$ Fructose-6-PO$_4$

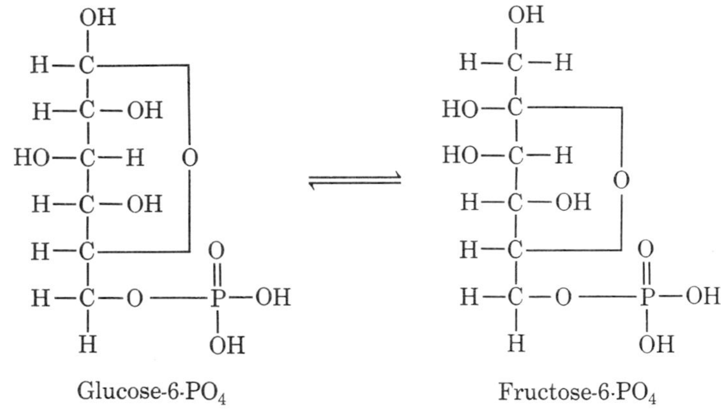

FIG. 8-30. Conversion of glucose-6-PO$_4$ to fructose-6-PO$_4$.

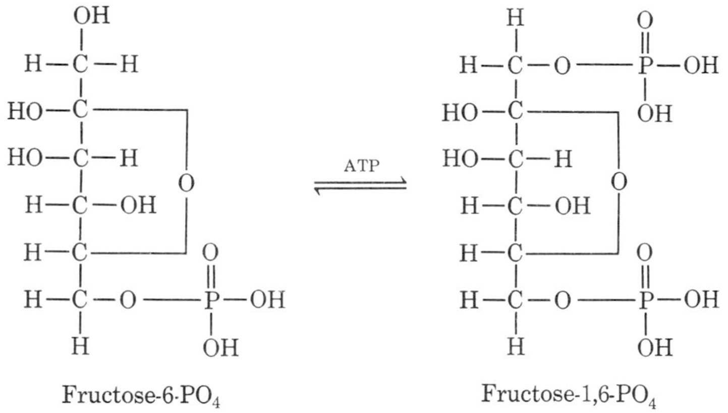

occurs, one cannot help but believe that the reaction has been oversimplified. Considering the mechanism of carbon splitting in other molecules, the reaction should be a split between two carbonyl groups (Fig. 8-33). With hydration of one of the three-carbon fragments it would be possible

Fructose-6-PO$_4$ Fructose-1,6-PO$_4$

FIG. 8-31. Phosphorylation of fructose-6-PO$_4$.

to form two molecules of 3-PO$_4$ glycerate. The metabolism of the 3-PO$_4$ glycerate could proceed as already shown in Fig. 8-16.

The second pattern of carbohydrate metabolism involves oxidation of carbon 1 to a carboxyl group (Fig. 8-34). The 6-PO$_4$ gluconate has a keto group the same as fructose. Actually, it would be possible to arrive at the same point starting at fructose-6-PO_4 (Fig. 8-35). Since this pattern of reactions follows the normal alcohol to aldehyde to acid metabolic pat-

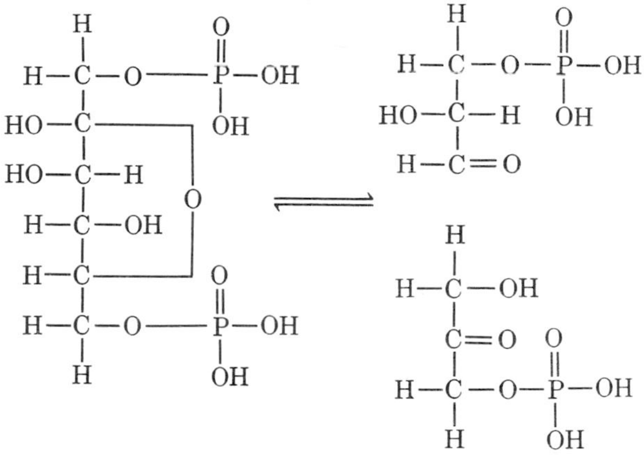

3-PO_4 glyceraldehyde

Dihydroxyacetone-PO_4

FIG. 8-32. Splitting fructose-1,6-PO_4 to two three-carbon fragments.

tern, one would expect the latter reaction to be shown to be the correct one. The keto group on carbon 2 permits decarboxylation to form a five-carbon sugar (Fig. 8-36).

Needless to say, the above reaction has been oversimplified. It is, in all

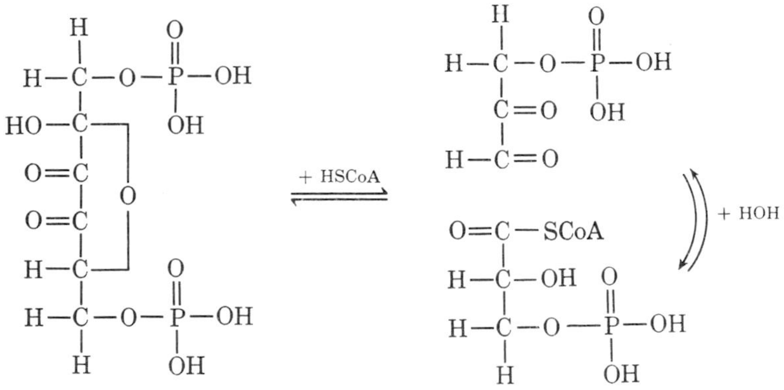

3-PO_4 glycerate

FIG. 8-33. Possible shear of fructose-1,6-PO_4.

probability, a multistep reaction of the same type shown for acids. The arabinose-5-PO_4 thus formed is converted to ribose-5-PO_4 for use in enzyme synthesis and in nucleoproteins.

Thus it is that hexoses are converted to pentoses. The question remains as to how the pentose is metabolized for energy. It is known that the pentose is split to a three-carbon fragment and a two-carbon fragment. If the split occurs as with the hexoses, $3\text{-}PO_4$ glyceric acid and PO_4 glycolic acid would be formed. These fragments would then be metabolized as shown previously.

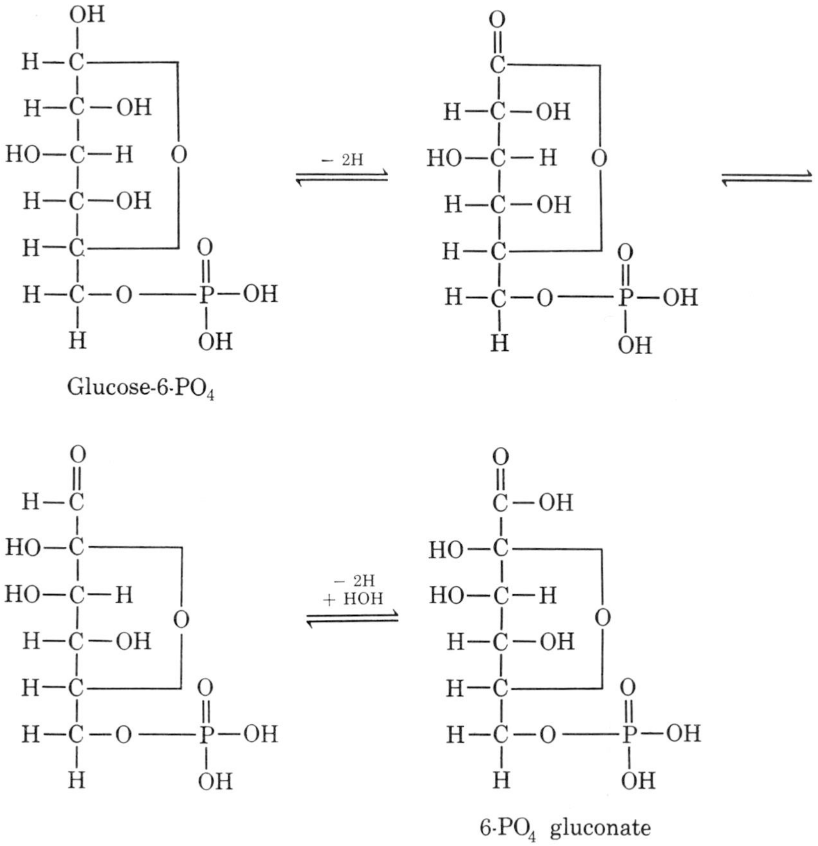

Glucose-6-PO_4

6-PO_4 gluconate

FIG. 8-34. Oxidation of glucose-6-PO_4 to 6-PO_4 gluconate.

Proteins

Proteins are the basic materials making up the majority of cellular protoplasm. For this reason microorganisms are able to break down and synthesize proteins very easily. Proteins are complex polymers of α-amino acids. Thus, the reactions of proteins are based on the reactions of acids which have already been shown and the reactions of the amino radical.

The reactions of the amino group have not been clearly defined. It has generally been stated that the amino group can be oxidized, reduced, hydrolyzed, or simply removed. Needless to say, such generalizations are designed to cover every possible reaction. It is the author's belief that the amino group undergoes one simple initial reaction followed by other reactions which have led to the rather vague generalizations for amino group reactions. The prime reaction at the amino groups appears to be hydrolysis (Fig. 8-37). The α-hydroxy acid can be oxidized to yield an

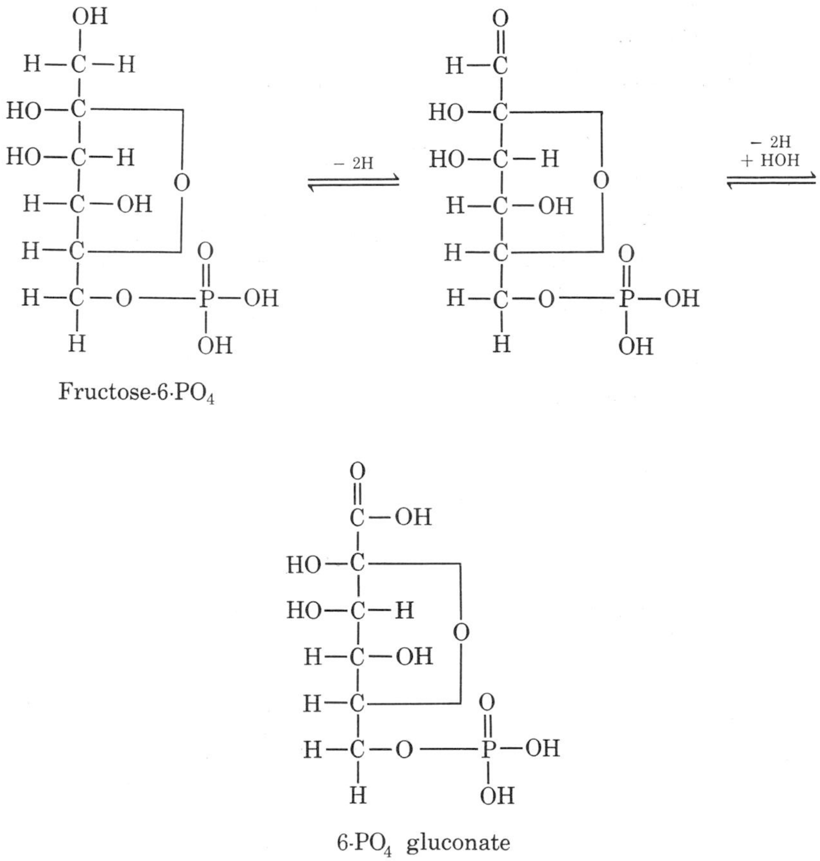

Fructose-6·PO$_4$

6·PO$_4$ gluconate

FIG. 8-35. Oxidation of fructose-6-PO$_4$ to 6-PO$_4$ gluconate.

α-keto acid (Fig. 8-38). The hydroxy acid can be dehydrated and then reduced to the saturated acid. It is easy to see from these reactions how the metabolic generalizations grew up.

In proteins the amino group of one amino acid reacts with the carboxyl group of another amino acid to form a peptide bond (Fig. 8-39). The ease of formation of the carbon-nitrogen bond greatly assists the micro-organisms in the formation and in the degradation of rather complex protoplasm structures such as the purines and pyrimidines.

The purines consist of adenine and guanine, while the pyrimidines consist of cyosine, uracil, and thymine, as shown in Fig. 8-40. It can be seen that the purines and pyrimidines have the same general structural core so that metabolism is similar in all compounds. The general pattern

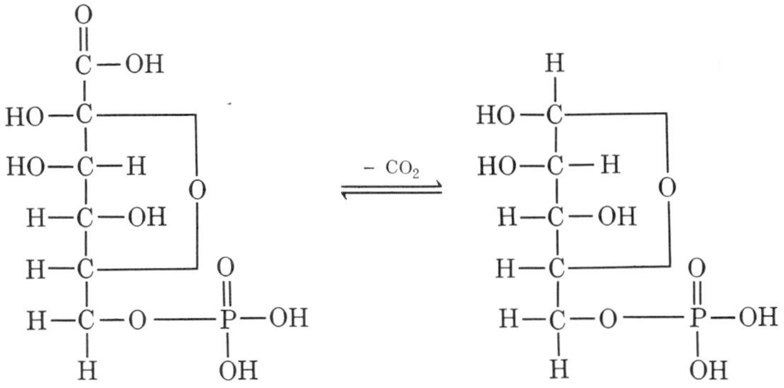

6·PO$_4$ gluconate   Arabinose-5-PO$_4$

FIG. 8-36. Decarboxylation of 6-PO$_4$ gluconate.

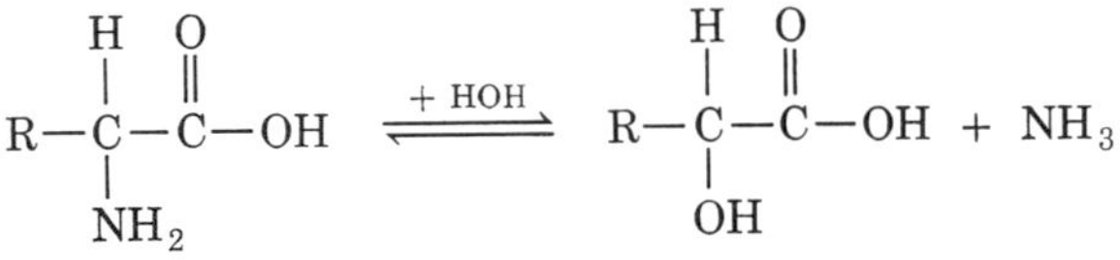

α-Amino acid   α-Hydroxy acid

FIG. 8-37. Hydrolysis of amino acids.

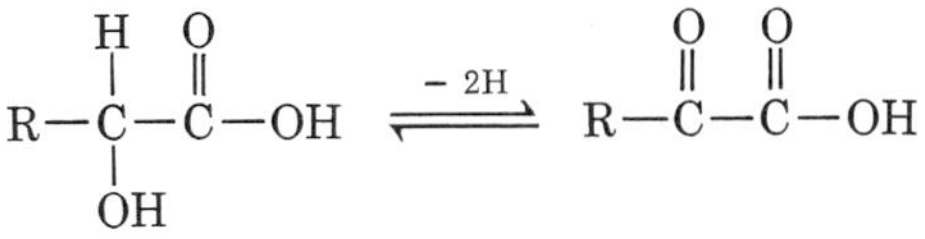

α-Hydroxy acid   α-Keto acid

(a) Oxidation

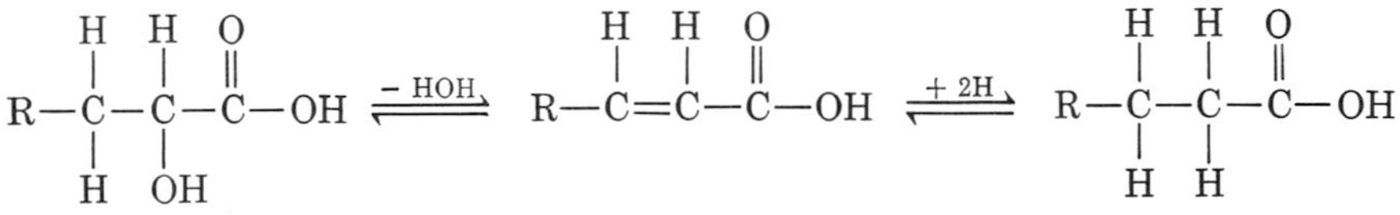

α-Hydroxy acid   Unsaturated acid   Saturated acid

(b) Dehydration and reduction

FIG. 8-38. Reactions of hydroxy acids.

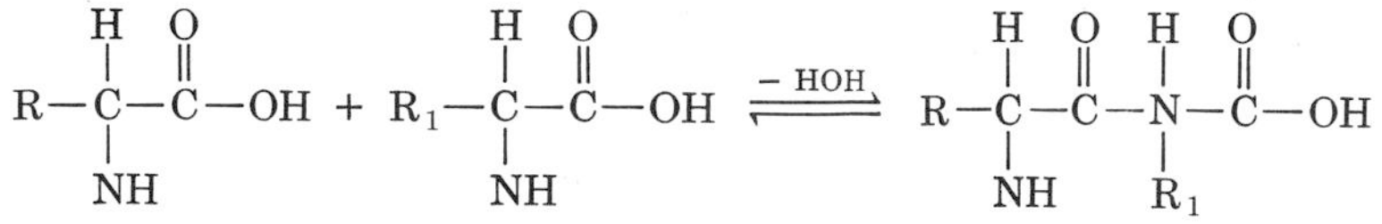

$$R-\underset{\underset{NH}{|}}{\overset{\overset{H}{|}}{C}}-\overset{\overset{O}{\|}}{C}-OH + R_1-\underset{\underset{NH}{|}}{\overset{\overset{H}{|}}{C}}-\overset{\overset{O}{\|}}{C}-OH \underset{}{\overset{-\ HOH}{\rightleftharpoons}} R-\underset{\underset{NH}{|}}{\overset{\overset{H}{|}}{C}}-\overset{\overset{O}{\|}}{C}-\underset{\underset{R_1}{|}}{\overset{\overset{H}{|}}{N}}-\overset{\overset{O}{\|}}{C}-OH$$

FIG. 8-39. Formation of peptide bond.

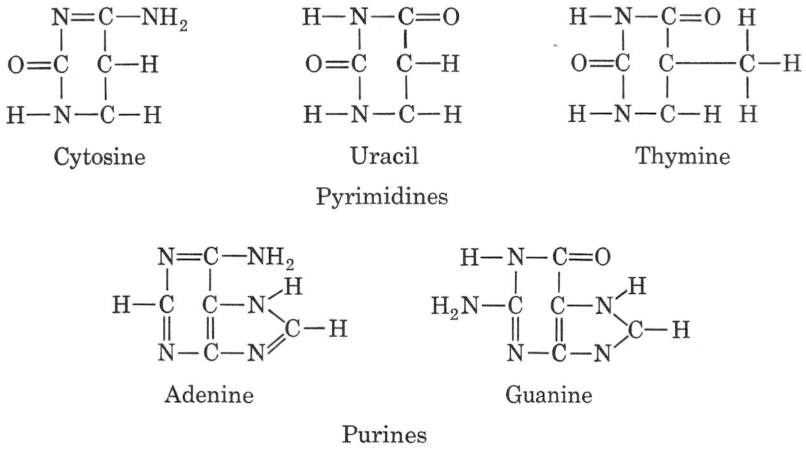

Cytosine	Uracil	Thymine

Pyrimidines

Adenine	Guanine

Purines

FIG. 8-40. Structure of purines and pyrimidines.

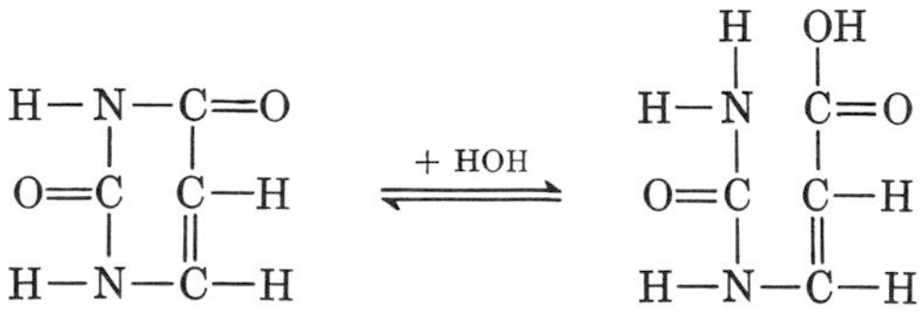

FIG. 8-41. Postulated hydrolysis of uracil.

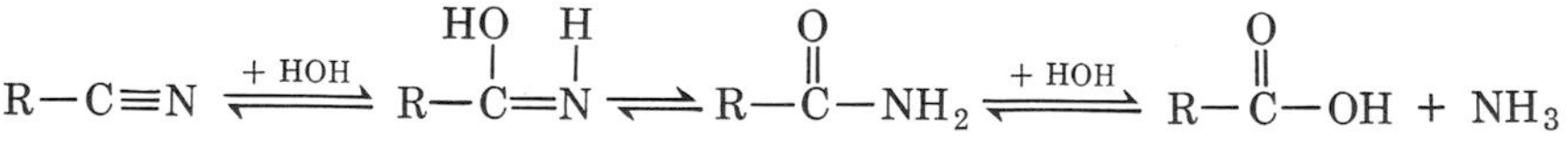

$$R-C\equiv N \underset{}{\overset{+\ HOH}{\rightleftharpoons}} R-\underset{}{\overset{\overset{HO\ \ H}{|\ \ \ |}}{C=N}} \rightleftharpoons R-\overset{\overset{O}{\|}}{C}-NH_2 \underset{}{\overset{+\ HOH}{\rightleftharpoons}} R-\overset{\overset{O}{\|}}{C}-OH + NH_3$$

FIG. 8-42. Metabolism of the cyano radical.

of reaction is related to the formation of peptide groups and the resulting hydrolysis to shear the carbon-nitrogen linkages (Fig. 8-41).

The amino group often appears on an organic molecule as a simple amine without the carboxyl group. The hydrolysis of the simple amine leads to the formation of a normal alcohol which can be readily metabolized by the patterns already shown.

Other Nitrogen Groups

Other nitrogen groups include the cyano group ($—C{\equiv}N$), the nitroso group ($—N{=}O$), and the nitro group ($—\overset{\displaystyle O}{\overset{\|}{N}}{=}O$). The cyano group is very reactive as a result of the triple bonds. This group takes up water very readily to form the amide group (Fig. 8-42). The amide is easily hydrolyzed to yield the acid and ammonia.

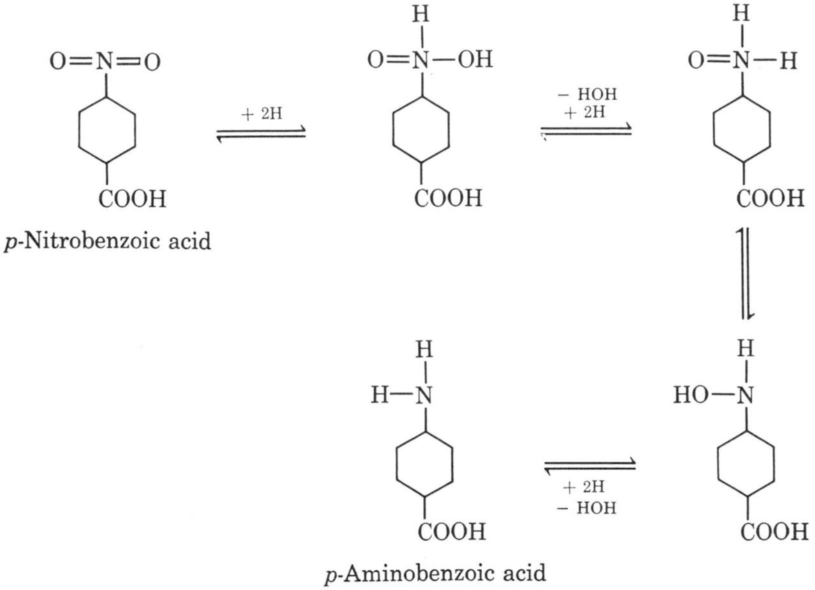

p-Nitrobenzoic acid

p-Aminobenzoic acid

FIG. 8-43. Postulated mechanism for the metabolism of p-nitrobenzoic acid to p-aminobenzoic acid.

The nitroso group and the nitro group have not been studied to any appreciable extent because of the extreme difficulty in metabolism of these compounds. The saturated nitrohydrocarbons are so volatile that they do not stay in solution long enough to be metabolized. The addition of other solubilizing groups such as a carboxyl group to nitrobenzene, forming p-nitrobenzoic acid, permits retention of the compound in solu-

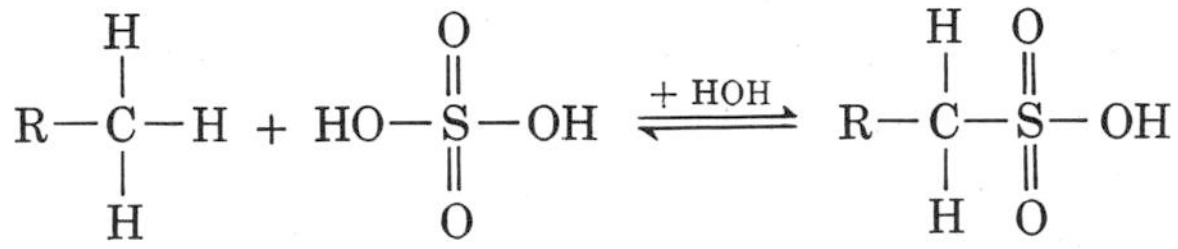

FIG. 8-44. Over-all sulfonation reaction.

tion long enough to be metabolized. Work with nitrobenzoic acid indicates reduction of the nitro group to an amino group prior to separation from the molecule (Fig. 8-43). The exact mechanism of this reduction has not been delineated to date.

Sulfur Groups

One of the least understood atoms in biochemical reactions is the sulfur atom. Part of the problem lies in the fact that sulfur does not occur to

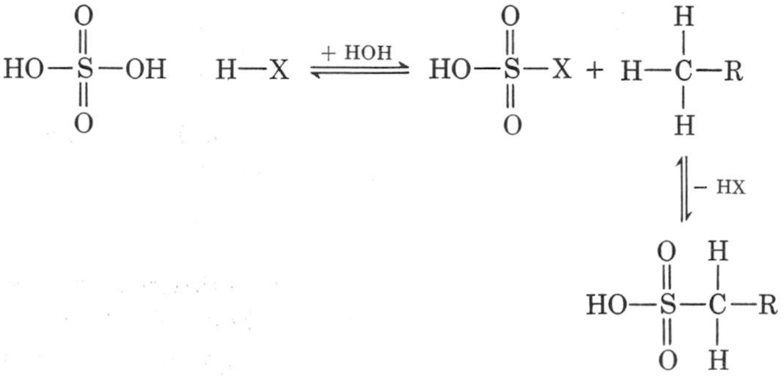

$$HO-\overset{\overset{O}{\|}}{\underset{\underset{O}{\|}}{S}}-OH \quad H-X \; \xrightleftharpoons{+ HOH} \; HO-\overset{\overset{O}{\|}}{\underset{\underset{O}{\|}}{S}}-X \; + \; H-\overset{\overset{H}{|}}{\underset{\underset{H}{|}}{C}}-R$$

$$\Big\downarrow {\scriptstyle - HX}$$

$$HO-\overset{\overset{O}{\|}}{\underset{\underset{O}{\|}}{S}}-\overset{\overset{H}{|}}{\underset{\underset{H}{|}}{C}}-R$$

Fig. 8-45. Possible sulfonation reaction.

an appreciable extent but is extremely important in reactions. The other part of the problem lies in the valence state of sulfur. Microorganisms have the ability to utilize sulfates as their source of sulfur. In sulfates sulfur has a valence of $+6$. But sulfur appears in the microbial cell as the sulfide with a valence of -2. Thus it is that all microorganisms utilizing sulfates as their source of sulfur have the ability to reduce sulfates; yet it is a highly specific reduction which can be used only in the synthesis reactions and not in the energy reactions which will be discussed in detail in the next chapter.

Sulfur winds up in cysteine, methionine, and CoA, as well as in other compounds of a lesser degree. The exact mechanism by which sulfur is incorporated into these compounds is not fully known, but it is believed to be related to a sulfonate formation (Fig. 8-44). The reaction shown is a gross oversimplification since the direct reaction is not possible. The initial reaction product has not been identified and is represented in Fig. 8-45. The sulfur

$$R-\overset{\overset{H}{|}}{\underset{\underset{H}{|}}{C}}-\overset{\overset{O}{\|}}{\underset{\underset{O}{\|}}{S}}-OH \; \xrightleftharpoons{+ 6H} \; R-\overset{\overset{H}{|}}{\underset{\underset{H}{|}}{C}}-SH$$

Sulfonate   Sulfide

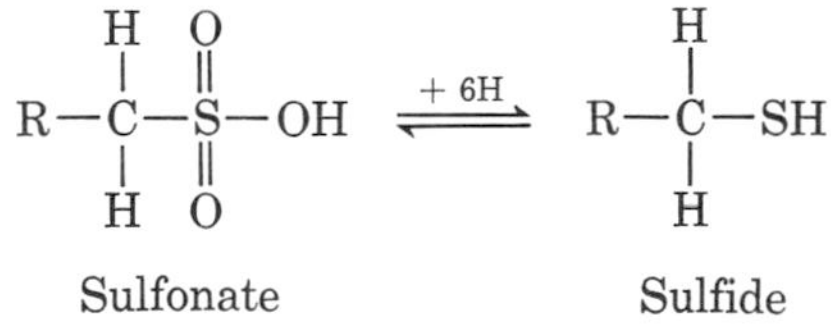

Fig. 8-46. Reduction of sulfonate to sulfide.

group is then transferred to its reactive compound and the unknown initial reactant is regenerated. Because of the valence changes in sulfur it is difficult to predict what intermediates occur during reduction to the sulfide. It suffices to give the over-all reaction (Fig. 8-46).

Actually the reaction in which sulfates are added to the hydrocarbon is the same type of reaction as the addition of carbon dioxide to a hydro-

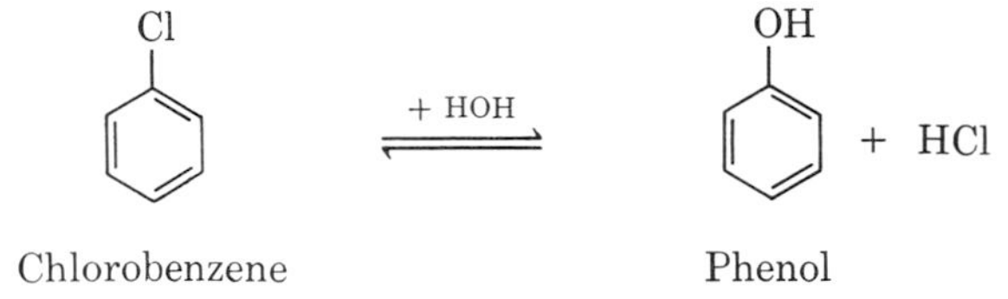

Chlorobenzene   Phenol

FIG. 8-47. Hydrolysis of chlorobenzene.

carbon. It may well be that CoA is involved in both reactions. If CoA is not involved, it is at least a CoA-type reaction.

Miscellaneous Groups

Other chemical groups of importance in industrial wastes include ethers, as well as chloro substituted hydrocarbons. While ethers do not occur to a large extent as such, the ether linkage is found in synthetic detergents made from ethylene oxide condensations.

$$-O-CH_2-CH_2-O-CH_2-CH_2-O-$$

While the metabolism of the ether linkage has not been elucidated, it is known that this linkage can be metabolized. It is postulated that the carbon atom adjacent to the oxygen atom is oxidized to a carbonyl group, thus forming an ester which is readily hydrolyzed to a carboxyl group and a hydroxyl group. The reaction is slow because it means an alpha oxidation rather than beta oxidation with the formation and metabolism of formic acid.

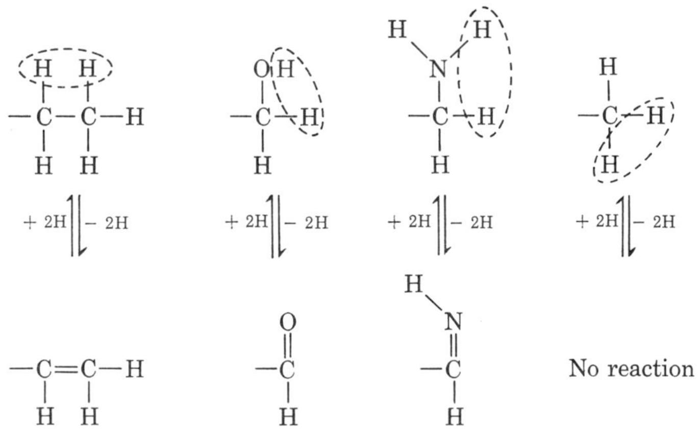

FIG. 8-48. Removal or addition of hydrogen.

Substitution of chloro groups for hydrogen has occurred extensively in many industrial compounds. Microorganisms have been shown to degrade many of the chloro compounds with the chloride ion as an end product of metabolism (Fig. 8-47). No one has studied how the chloro group is removed.

General Metabolic Reactions

For the most part, metabolism has been studied as a series of reactions related only to one starting material. It is my opinion that while this approach is necessary, it has led to much confusion and delay in understanding metabolism. The metabolic reactions occurring within the microbial cell follow definite patterns which are all interrelated. While certain specific microorganisms have unusual

$$
\underset{\overset{|}{H}}{\overset{\overset{|}{H}}{-C}}-\underset{\overset{|}{H}}{\overset{\overset{|}{OH}}{C}}-\ \underset{+ HOH}{\overset{- HOH}{\rightleftharpoons}}\ \underset{\overset{|}{H}}{\overset{}{-C}}=\underset{\overset{|}{H}}{\overset{}{C}}-
$$

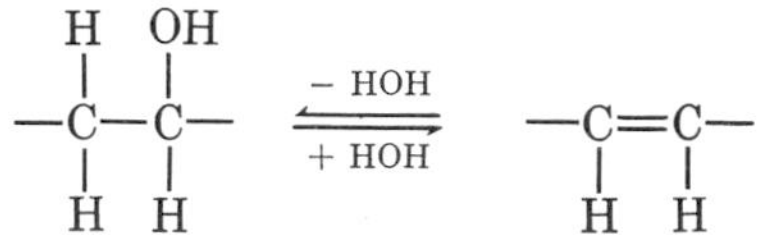

Fig. 8-49. Removal or addition of water.

metabolic pathways with exotic reactions, the microorganisms responsible for stabilizing organic matter in nature have simple metabolic patterns designed for complete metabolism. They do not produce strange and exotic chemical reactions. Their patterns of metabolism are the same for all organic compounds.

Summarizing the general metabolic reactions we find only a few key reactions. With an understanding of these metabolic reactions, it is possi-

$$
\underset{}{\overset{O\ \ O}{\overset{\|\ \ \|}{-C-C-}}}\ \underset{- HSCoA}{\overset{+ HSCoA}{\rightleftharpoons}}\ \overset{O}{\overset{\|}{-CSCoA}} + \overset{O}{\overset{\|}{H-C}}
$$

$$
\underset{\overset{|}{H}}{\overset{O\ \ H\ \ O}{\overset{\|\ \ |\ \ \|}{-C-C-C}}}-SCoA\ \underset{- HSCoA}{\overset{+ HSCoA}{\rightleftharpoons}}\ \overset{O}{\overset{\|}{-C}}-SCoA + \underset{\overset{|}{H}}{\overset{H\ \ O}{\overset{|\ \ \|}{H-C-C}}}-SCoA
$$

$$
\underset{H-\overset{|}{C}-H\ \ H}{\overset{H\ \ \ \ \ \ H\ \ O}{\overset{|\ \ \ \ \ \ |\ \ \|}{-C——C-C}}}-SCoA\ \rightleftharpoons\ \underset{H-\overset{\|}{C}}{\overset{H}{\overset{|}{-C}}} + \underset{\overset{|}{H}}{\overset{H\ \ O}{\overset{|\ \ \|}{H-C-C}}}-SCoA
$$

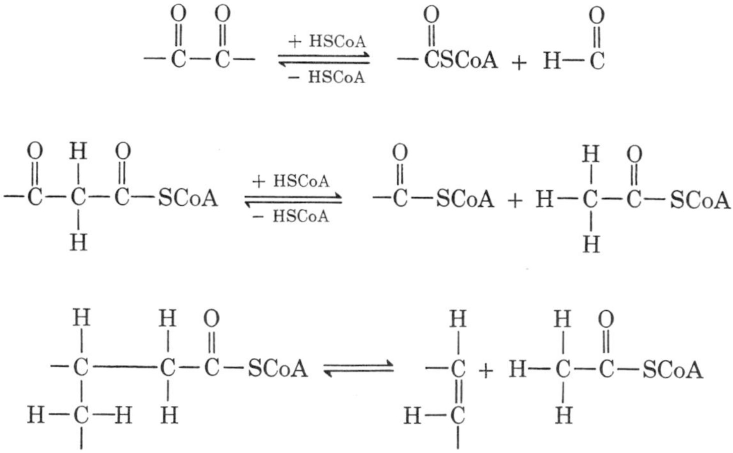

Fig. 8-50. Carbon-carbon split or formation.

ble to postulate the degradation of almost any type of organic compound found in wastes.

1. *Addition or removal of hydrogen.* Hydrogen atoms are always added or removed in pairs from adjacent atoms, never from the same atom (Fig. 8-48).

2. *Addition or removal of water.* Water can be added to adjacent atoms having an unsaturated bond between the atoms or removed from adjacent atoms to create an unsaturated bond (Fig. 8-49).

3. *Carbon-carbon split or formation.* Carbon-carbon bond splits occur easiest between a dicarbonyl group but can occur at certain monocarbonyl groups with CoA. Acetyl- and formyl-CoA fragments may be removed from complex molecules provided the residual fraction is not insoluble. The carbon-carbon bond formation occurs primarily as a reverse of the carbon-carbon split but can also occur between an unsaturated bond (Fig. 8-50).

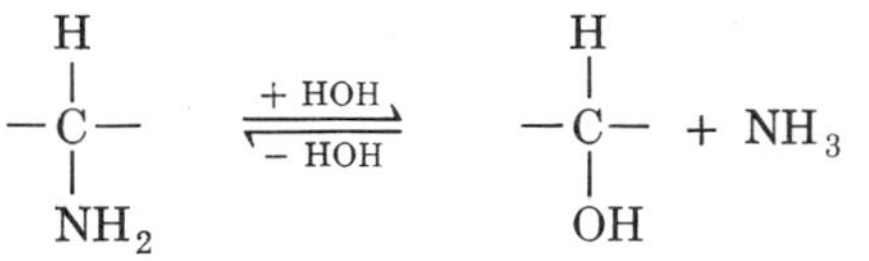

$$-\overset{\displaystyle H}{\underset{\displaystyle NH_2}{C}}- \quad \underset{- HOH}{\overset{+ HOH}{\rightleftharpoons}} \quad -\overset{\displaystyle H}{\underset{\displaystyle OH}{C}}- \ + \ NH_3$$

FIG. 8-51. Addition or removal of nitrogen.

4. *Addition or removal of nitrogen.* Nitrogen appears to be added or removed with a hydrolysis reaction as ammonia. The ammonia displaces water from the molecule upon addition and is displaced by water upon removal (Fig. 8-51).

It may seem to many microbiologists that this is a gross oversimplification of a complex series of reactions. It is, but it does enable the microbiologists to predict the rough pattern of metabolism and to arrive at the end product of metabolism correctly. To this end these simple rules have value.

SUGGESTED REFERENCE

1. Umbreit, W. W., "Metabolic Maps," Burgess Publishing Company, Minneapolis, Minn., 1952.

CHAPTER 9

Energy

The growth and survival of microorganisms depends upon their ability to obtain energy from the system. Energy is required for the production of new protoplasm, for motility, and just to remain alive. Microorganisms obtain their energy from the metabolism of organic and inorganic compounds. The metabolic reactions have already been delineated, but the question to be answered is how do the microorganisms obtain energy from these reactions?

ENERGY TRANSFER

One of the most difficult problems in microbial physiology has been the determination of the energy yield to the microorganisms during metabolism. Efforts to use the conventional concepts of thermodynamics have not proved at all satisfactory. In thermodynamics the oxidation of organic or inorganic compounds releases heat energy, but microorganisms are not heat engines and cannot utilize heat energy. As a result, microorganisms are forced to prevent the loss of the chemical energy in the form of heat. This is done with coupled chemical reactions in which energy is given up from one compound to another compound with only a very minor heat loss.

The currently accepted biological scheme of energy change involves the use of a phosphate enzyme system. The coenzymes, ADP and ATP, are high-energy compounds in that they contain high-energy phosphate bondings. As the chemical reactions release energy, inorganic phosphate is added to ADP to form ATP. In this way the energy is stored in the ATP rather than lost as heat. As the microorganisms require the energy, the ATP is reduced back to ADP with a transfer of energy to the chemical reaction needing it.

The net result of the energy process is shown in Fig. 9-1. The energy level of the organic matter being metabolized has decreased while the energy level of cellular material has risen. The rise in energy level of the cellular material is not so high as the decrease in energy of the metabolized material. This results from heat energy lost from the system during the step reaction. Since the heat energy lost is a function of the reactions

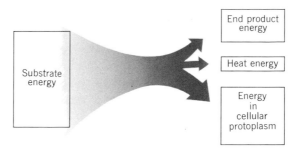

FIG. 9-1. Energy changes during metabolism.

and hence a function of energy transferred, heat energy can be used as a measure of energy transfer.

$$\text{Biological energy} = K \, \Delta \, \text{heat energy} \qquad (9\text{-}1)$$

Since the number of reactions requiring energy within a microorganism is very large, the change in biological energy is directly proportional to the change in heat energy as shown in Eq. (9-1). In view of this concept the question arises as to why the conventional concepts of thermodynamics have not proved satisfactory. The prime reason for this has been the failure to make a complete energy balance within the system to determine what the thermodynamic relationships really were.

OXIDATION

Energy is released by oxidation reactions. Since organic matter in waste waters is stabilized by oxidation, it is essential that the oxidation reactions be well understood. Bacteria and other microorganisms in waste stabilization systems do not oxidize matter by the direct addition of oxygen, but rather by the indirect scheme of hydrogen removal and addition of water as shown in Figs. 8-48 and 8-49. The hydrogen eventually reacts with oxygen, carbon, nitrogen, or sulfur.

The chemical scheme of oxidation is believed to be the same for all microorganisms whether plant or animal. The differences between aerobic, facultative, and anaerobic bacteria lie in their mechanisms of hydrogen oxidation. Strict aerobes utilize "free," dissolved oxygen for their ultimate hydrogen acceptor, while strict anaerobes use "chemically bound" oxygen, carbon, nitrogen, or sulfur as their hydrogen acceptor. The facultative bacteria can use most of the above mechanisms, but will always use the hydrogen acceptor yielding the greatest energy. Thus, facultative bacteria will not use carbon as their hydrogen acceptor when dissolved oxygen is present.

HYDROGEN REMOVAL

The removal of hydrogen from organic compounds follows a definite pattern, as already indicated in Fig. 8-47. The hydrogen removal is brought about by the coenzymes, DPN or TPN. In both DPN and TPN the key to hydrogen transfer lies in the valence change of nitrogen in the nicotinic acid fraction of the coenzymes. In the oxidized state nitrogen has a valence of 5. Removal of two hydrogen atoms results in a change in valence of nitrogen to 3. Actually, the removal of two hydrogen atoms yields two electrons and two hydrogen ions

$$2H \rightarrow 2H^+ + 2e \qquad (9\text{-}2)$$

DPN and TPN take up one hydrogen ion and two electrons, leaving one

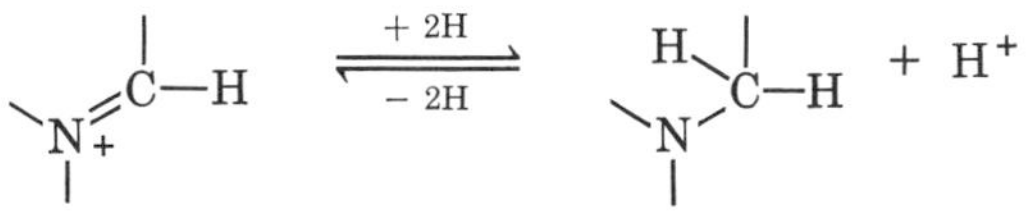

Fig. 9-2. Oxidation-reduction mechanism of DPN.

hydrogen ion free. The reduced form of DPN is written DPNH or $DPNH_2$; the latter form is used in this text (Fig. 9-3).

DPNH$_2$ REGENERATION

The quantity of DPN within the cell is definitely limited. If all the DPN were reduced to $DPNH_2$, the hydrogen transfer reaction would cease. Thus, it is necessary to regenerate continuously the $DPNH_2$. The difference between aerobic and anaerobic metabolism lies in the method of $DPNH_2$ regeneration.

Aerobic Metabolism

In aerobic microorganisms the $DPNH_2$ is regenerated by flavin adenosine dinucleotide, FAD, which goes to $FADH_2$.

$$DPNH_2 + FAD \rightarrow DPN + FADH_2 \qquad (9\text{-}3)$$

The $FADH_2$ must also be regenerated if the reaction is to proceed. Cytochrome pigments regenerate the $FADH_2$ and in turn react with molecular oxygen to form water, the end product of hydrogen oxidation.

Cytochrome Pigments. There are three different cytochrome pigments, *b, c,* and *a.* Each type of pigment has several different modifications so that there are many more than three cytochromes. While all the cyto-chrome pigments act similarly, not all cytochromes exist in all micro-organisms. Basically the cytochromes utilize the oxidation and reduction of iron for hydrogen transfer. This reaction is strictly an electron transfer.

$$FADH_2 + 2Fe^{+3} \rightarrow FAD + 2Fe^{++} + H^+ \tag{9-4}$$

The cytochromes are regenerated by a series transfer reaction in which one pigment is regenerated by reducing the next pigment. The final reaction occurs between dissolved oxygen and the two hydrogen ions liberated in the earlier reactions.

$$\tfrac{1}{2}O_2 + 2H^+ + 2Fe^{++} \rightarrow 2Fe^{+3} + H_2O \tag{9-5}$$

The complete scheme of aerobic hydrogen transfer is shown in Fig. 9-3.

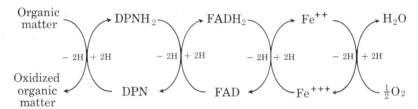

FIG. 9-3. Scheme of aerobic hydrogen transfer.

Anaerobic Metabolism

The strict anaerobes normally lack the cytochrome pigments and must regenerate their $DPNH_2$ directly or through FAD. There are three groups of strict anaerobes: (1) the sulfate reducers, (2) the methane formers, and (3) the organic reducers.

The sulfate-reducing bacteria are a highly specialized group of bacteria, *Desulfovibrio,* which utilize sulfates as their hydrogen acceptor. This group of bacteria is very important to sanitary engineers in sewer-pipe corrosion and in odor production. Yet, little is known about these bacteria except that sulfates are reduced to sulfides. The reduction of sulfates probably occurs in a stepwise series of reactions in which a cytochrome-type system is involved. The postulated reactions are shown in Fig. 9-4. One of the most interesting aspects of the sulfate-reducing bacteria is

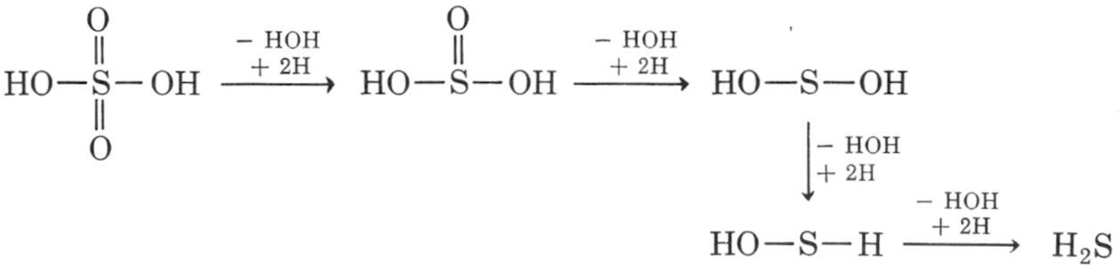

$$HO-\overset{\overset{\displaystyle O}{\|}}{\underset{\underset{\displaystyle O}{\|}}{S}}-OH \xrightarrow[+2H]{-HOH} HO-\overset{\overset{\displaystyle O}{\|}}{S}-OH \xrightarrow[+2H]{-HOH} HO-S-OH$$

$$\downarrow {\scriptstyle -HOH \atop +2H}$$

$$HO-S-H \xrightarrow[+2H]{-HOH} H_2S$$

FIG. 9-4. Possible mechanism of sulfate reduction.

the highly specific character of these bacteria, especially in view of the fact that most of the common bacteria reduce sulfates to sulfide in their protoplasm. The sulfate reducers must have the sulfate tied up with an unusual enzyme group or must have an unusual hydrogen-transfer system as they are the only bacteria capable of using sulfates as a hydrogen acceptor in their energy metabolism.

The methane-producing bacteria are like the sulfate reducers in their highly specific characteristics. Their ability to utilize carbon for their only hydrogen acceptor is unique in microbiology. Very little is known about the exact mechanism of how methane is formed. As already indicated in Fig. 8-1, it is produced from the reduction of carbon dioxide. Yet, the question arises as to where the hydrogen comes from. One thing for certain, it does not come from the oxidation of carbohydrates or proteins. It appears that most of the methane is formed from short-chain fatty acids such as formic and acetic.

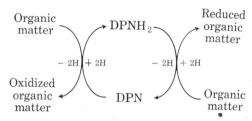

FIG. 9-5. Hydrogen transfer by anaerobic bacteria.

The reduction of formic acid occurs with the formation of carbon dioxide as methane is formed. Like all biochemical reactions this reaction is probably not direct, but rather is a complex reaction. The use of acetic acid for methane formation has been sufficiently studied to know that carbon 1 winds up as carbon dioxide while carbon 2 winds up as methane. The reaction is not a direct shearing, but probably involves a split to two single carbon complexes, with one complex reducing the other complex.

The majority of anaerobic bacteria regenerate their $DPNH_2$ by direct reaction with oxygen containing organic matter (Fig. 9-5). Carbohydrates contain considerable chemical oxygen and are good hydrogen acceptors. On the other end of the scale are the long-chain saturated alcohols which are almost inert anaerobically. The lack of hydrogen acceptor is the chief reason for the inability of anaerobic bacteria to completely degrade organic matter. The end products of anaerobic metabolism are acids, aldehydes, ketones, and alcohols. The balance of end products is the result of $DPNH_2$ regeneration.

The facultative bacteria can utilize nitrites and nitrates as their hydrogen acceptors much like the sulfate reducers utilize sulfate. The nitrite-nitrate reduction scheme is tied in with the cytochrome system. Although the entire scheme of nitrate reduction has not been observed, it is known that nitrates are reduced to nitrites and to some unidentified intermediates before being fully reduced to nitrogen gas. A postulated pattern of nitrate reduction is given in Fig. 9-6. Note once again the use of an

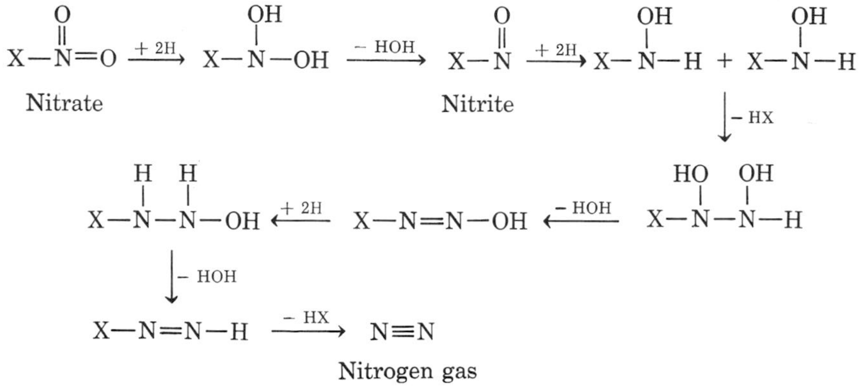

$$X-\overset{\overset{\displaystyle O}{\|}}{N}=O \xrightarrow{+2H} X-\overset{\overset{\displaystyle OH}{|}}{N}-OH \xrightarrow{-HOH} X-\overset{\overset{\displaystyle O}{\|}}{N} \xrightarrow{+2H} X-\overset{\overset{\displaystyle OH}{|}}{N}-H + X-\overset{\overset{\displaystyle OH}{|}}{N}-H$$

Nitrate   Nitrite $\downarrow -HX$

$$X-\overset{\overset{\displaystyle H}{|}}{N}-\overset{\overset{\displaystyle H}{|}}{N}-OH \xleftarrow{+2H} X-N=N-OH \xleftarrow{-HOH} X-\overset{\overset{\displaystyle HO}{|}}{N}-\overset{\overset{\displaystyle OH}{|}}{N}-H$$

$\downarrow -HOH$

$$X-N=N-H \xrightarrow{-HX} N\equiv N$$

Nitrogen gas

FIG. 9-6. Postulated mechanism of nitrate reduction.

unknown complexing agent similar to CoA. In all probability there is a strong tie because of the acidic character of the molecules. The use of nitrates in this manner for a hydrogen acceptor is known as *denitrification*.

Energy Exchange. The removal of two hydrogen atoms causes a rupture in bonds releasing energy, but a portion of that energy is used in creating new bonds in the DPN molecule. Part of the energy released appears to go into the formation of the so-called "high-energy" phosphate bonds in which inorganic phosphorus is added to adenosine diphosphate, ADP, to form ATP (Fig. 9-7). The remainder of the energy is lost as heat.

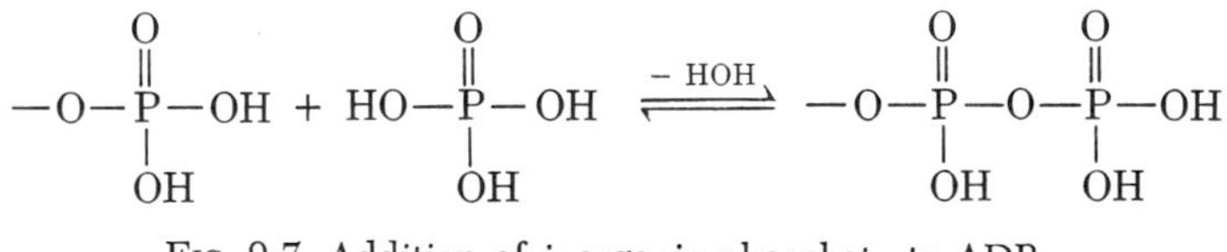

$$-O-\overset{\overset{\displaystyle O}{\|}}{\underset{\underset{\displaystyle OH}{|}}{P}}-OH + HO-\overset{\overset{\displaystyle O}{\|}}{\underset{\underset{\displaystyle OH}{|}}{P}}-OH \overset{-HOH}{\rightleftharpoons} -O-\overset{\overset{\displaystyle O}{\|}}{\underset{\underset{\displaystyle OH}{|}}{P}}-O-\overset{\overset{\displaystyle O}{\|}}{\underset{\underset{\displaystyle OH}{|}}{P}}-OH$$

FIG. 9-7. Addition of inorganic phosphate to ADP.

When DPNH$_2$ is regenerated by FAD, there is another formation of ATP from ADP. This is repeated again when the FADH$_2$ is regenerated in the cytochrome system and again when the cytochrome pigments are regenerated. Thus, the energy released during the oxidation of hydrogen is largely trapped in ATP.

The energy in the ATP is used in transferring phosphorus to other molecules and reducing ATP to ADP. The resulting organic phosphate is eventually hydrolyzed to inorganic phosphate with little energy loss. Thus it is that phosphate is merely an intermediary in promoting reactions by supplying sufficient energy to drive the reaction.

Energy within the cell is used primarily for synthesis. The cell is continuously synthesizing new cellular components to replace displaced or worn structures. A portion of the hydrogen removed by DPN is used in synthesis of reduced structures. The ATP is broken to ADP whenever $DPNH_2$ gives up its hydrogen to another organic molecule. It is obvious that more ATP molecules will be created than degraded by this scheme. The excess ATP molecules will be used to drive synthesis reactions other than the addition of hydrogen to organic molecules.

The splitting of carbon-carbon bonds yields energy which also is used to convert ADP to ATP. Much of this energy is used to build up protoplasmic structures. It is doubtful if ATP is formed when hydrolysis occurs. The energy released on hydrolysis is very low compared to the energy released on splitting carbon-carbon bonds or on removal of hydrogen. Yet, the removal of water from a molecule during synthesis requires energy. The addition of ammonia to form amino acids requires energy, and yet it is doubtful if the release of ammonia during metabolism of amino acids yields energy to the cell, since there is not enough energy released to form a high-energy bond.

Autotrophic Metabolism

While most microorganisms obtain their energy from the oxidation of organic matter, the autotrophic bacteria obtain their energy from the oxidation of inorganic compounds. The nitrifying bacteria utilize ammonia and nitrite as their energy source. Sulfur bacteria use hydrogen sulfide and other reduced forms of sulfur. Iron bacteria oxidize ferrous iron while the hydrogen bacteria oxidize hydrogen.

If little is known about the energy reactions of the heterotrophic bacteria, even less is known about the autotrophic bacteria. The gross energy reaction for the autotrophs are known, but the exact energy pattern is far from known.

Nitrifying Bacteria. The nitrifying bacteria have the ability to convert ammonia to nitrates. The reactions occur in two phases. In phase one the *Nitroso-* bacteria convert ammonia to nitrite. The reaction has been postulated in Fig. 9-8. The 3 DPN molecules reduced yield considerable energy to the *Nitroso-* bacteria.

The *Nitro-* bacteria take the nitrites and oxidize them to nitrates (Fig. 9-9). Only one DPN is produced by this reaction. The low-energy yield requires the *Nitro-* bacteria to process three times as much substrate as the *Nitroso-* to obtain the same energy. This helps to explain the rapid conversion of nitrite to nitrate in mixed populations. The demand for energy prevents a large nitrite build-up. The DPN is regenerated by dissolved oxygen in the same way as heterotrophic bacteria.

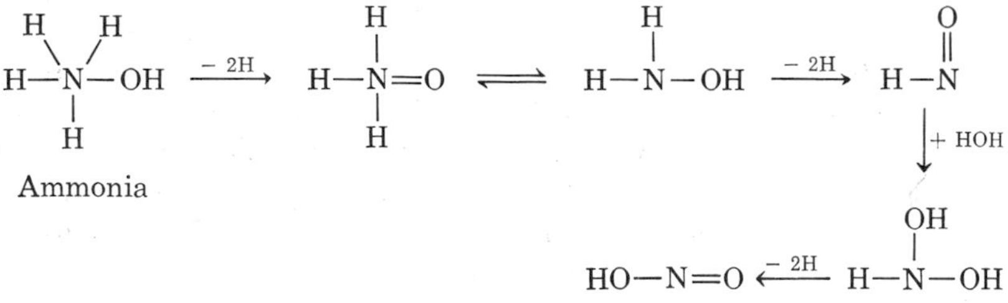

Fig. 9-8. Oxidation of ammonia to nitrite.

Sulfur Bacteria. The sulfur bacteria obtain their energy from the oxidation of hydrogen sulfide, thiosulfate, tetrathionate, or other reduced sulfur compounds. The most interesting aspect of the sulfur bacteria, *Thiobacillus,* is their ability to survive at pH values of 1.0 and under. The pattern of hydrogen sulfide oxidation is about as well known as the pattern of sulfate reduction. The starting material and the end product are known but there is a lot of vacant space in between these two points. It is postulated as the reverse of sulfate reduction (Fig. 9-4).

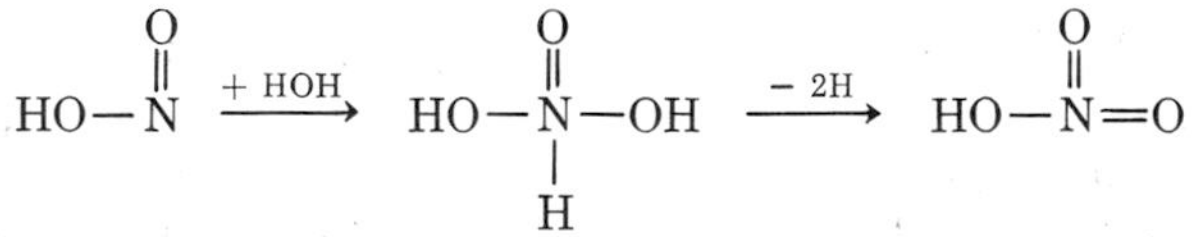

Fig. 9-9. Oxidation of nitrite to nitrate.

Iron Bacteria. The ability of bacteria to grow from the energy obtained from the oxidation of ferrous iron to ferric iron poses an interesting scheme in metabolism. Most of these organisms have an iron oxide coating around the cell. The ferrous iron is soluble and is readily taken into the cell. The removal of an electron results in formation of ferric iron which readily forms insoluble oxides. As long as the ferric iron is within the cell, it is complexed with organic compounds to prevent precipitation. As the ferric iron is released from the cell, the insoluble oxide forms a coating around the cell.

Hydrogen Bacteria. *Hydrogenomonas* are bacteria which can oxidize hydrogen gas to water and obtain their energy. The simplicity of these microorganisms is apparent from the single energy-yielding reaction.

Energy Transfer. The autotrophic bacteria are all strict aerobes and probably have as complete a hydrogen-oxidizing system as the aerobic, heterotrophic bacteria. One of the most intriguing questions is why the *Nitroso-* bacteria oxidize ammonia only to nitrites and the *Nitro-* bacteria are unable to oxidize ammonia. These phenomena can only occur if the *Nitroso-* bacteria lack a portion of the cytochrome system, namely, cytochrome a. The *Nitro-* bacteria on the other hand must depend entirely on the cytochromes for their hydrogen-transfer system.

Energy Yield. The survival of any microorganism depends upon its ability to obtain sufficient energy for its normal life functions. If each group of organisms metabolized 1 gm of its energy material, the approximate energy yields would look like this:

Nitroso- (ammonia → nitrite)................... 2 kcal/gm
Nitro- (nitrite → nitrate)................ .... 0.3 kcal/gm
Thiobacillus (sulfide → sulfate)................. 5 kcal/gm
Iron-oxidizing bacteria (Fe^{++} → Fe^{+3})........... 0.2 kcal/gm
Hydrogenomonas (hydrogen → water)...:........ 28 kcal/gm

The hydrogen-oxidizing bacteria obtain considerably more energy than any of the other bacteria. The iron-oxidizing bacteria must process considerably more material than the hydrogen-oxidizing bacteria to obtain the same amount of energy. The importance of this becomes readily apparent when considering the synthesis of new cells.

Photosynthesis. Algae contain chlorophyll which permits them to obtain their energy directly from the sun's rays. This process permits the algae to grow in a completely inorganic environment without oxidizing any material for energy. The photosynthetic pigments supply the energy for the dehydrogenation of water by triphosphopyridine nucleotide (TPN).

$$TPN + 2HOH \xrightarrow[\text{Chlorophyll}]{\text{Sunlight}} TPNH_2 + H_2O_2$$
$$\downarrow$$
$$\tfrac{1}{2}O_2 + H_2O \qquad (9\text{-}6)$$

In dehydrogenating water, the algae follow the general metabolic pattern of removing the hydrogen atoms from adjacent atoms and not from the same atom. In water this means removal from two molecules of water. Actually, the dimer characteristics of water permit this transfer more easily than would be initially expected (Fig. 9-10). The two end hydrogen atoms are removed by TPN, forming a peroxide which is broken down to yield water and oxygen. Thus it is that the oxygen produced

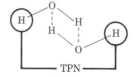

FIG. 9-10. Removal of hydrogen from water by algae.

by the algae comes from water. As will be shown in the next chapter, the TPNH$_2$ is regenerated in synthesis of new cells. Needless to say, a high-energy phosphate conversion also takes place during the removal of hydrogen.

$$ADP + inorg. \ PO_4 \rightarrow ATP \tag{9-7}$$

BIOLOGICAL ENERGY

The availability or lack of availability of energy to the microbial cell has long puzzled the microbiologists. The conventional concepts of energy are all based on the thermodynamics of heat engines. But microorganisms are not heat engines and cannot utilize the heat released from a reaction. Then what do microorganisms have to do with conventional thermodynamics?

Actually, microorganisms follow the laws of thermodynamics as well as any heat engine. The chemical reactions which occur within a cell either yield energy or require energy. We have already seen that the microorganisms obtain and utilize energy through a series of coupled chemical reactions in which there is always a decrease in energy level. This is true even in the synthesis reactions. There is also always a portion of the energy lost as waste heat the same as there is for any chemical reaction.

Heat Energy and Free Energy

The oxidation of any compound yields a certain quantity of heat energy. This heat energy yielded was the basis for thermodynamics and is known as the heat of combustion when the compound is oxidized completely. Each compound also has a certain energy level known as the heat of formation which is the heat energy required to form the compound from its elements. All elements are arbitrarily set at zero energy level.

Since microorganisms could not utilize heat energy, the microbiologists did not like to use the heats of formation and the heats of combustion to express energy change. Instead they turned to the thermodynamic concept of free energy which is a measure of the actual energy available for work. The change in free energy is expressed by Eq. (9-8).

$$\Delta F = \Delta H - T \, \Delta S \tag{9-8}$$

where F = free-energy change
H = heat-energy change
S = entropy change
T = absolute temperature

The entropy S is that energy that is not available for work and hence is lost.

But even free energy has not worked out to explain microbial metabolism with conversion of energy into protoplasm. The reason for this lies in the fact that all the free energy in a compound is not available for microbial use. Failure of the microorganism to convert all the energy from a reaction results in its loss.

Actually, both heat energy and free energy are a function of the *biological energy*, the energy available to the microorganisms, which in turn is a function of the chemical compound being metabolized. Unfortunately, so little quantitative work has been done on energetics that it is not possible to indicate the exact relationship between all three forms of energy. The best criteria for the expenditure of biological energy is in the oxygen utilized under aerobic conditions. The utilization of oxygen is for the regeneration of $DPNH_2$, but carbon is oxidized indirectly by the addition of water with the subsequent removal of hydrogen by DPN. The oxygen uptake then is a function of both the hydrogen removal and carbon oxidation, the major two sources of biological energy.

Warburg Respirometer

The Warburg respirometer is the primary apparatus for the determination of oxygen utilization. A Warburg respirometer is shown in Fig. 9-11. It consists of a controlled temperature water bath in which the sample is placed. The sample is contained in a flask attached to a manometer. The manometer-flask unit is mounted on a shaking device so that the sample is continuously agitated.

The flask is attached to the closed leg of the manometer and the manometer adjusted to give a constant gas volume over the sample. The water bath keeps the temperature constant so that any gaseous interchange between the gas over the sample and the sample results in a change in gas pressure and hence a change on the manometer open leg. If a solution of 10 per cent KOH is used in the center well of the flask and a known volume of sample is used, it is possible to calibrate the manometer to yield oxygen uptake in milligrams per liter.

In sanitary microbiology, two types of Warburg flasks are used, the 25-ml side-arm flask and the 125-ml B.O.D. flask. The small side-arm flask permits the addition of substrate at any time after the Warburg run has started. The large flask has the advantage of a larger sample and hence less sampling error.

The major advantage in the Warburg lies in the fact that continuous readings in oxygen uptake are permissible for any time period. Readings can be spaced from 5-min intervals to several hours, depending upon the rate of oxygen uptake. It is possible to determine the carbon dioxide produced during oxygen uptake by the use of two identical samples, one with KOH and one without KOH. The sample with KOH in the center well measures only the oxygen uptake, while the sample without the KOH measures the pressure change due to oxygen uptake and carbon dioxide released. Unfortunately, the release of carbon dioxide permits bicarbonate formation in the sample and does not allow complete release to the gaseous phase of carbon dioxide. This problem can be reduced to a minimum by buffering the pH of the sample at approximately 6.0. From the partial pressure of carbon dioxide and the pH it is possible to cal-

FIG. 9-11. Warburg respirometer.

culate the quantity of carbon dioxide tied up in solution as bicarbonate. The correction for carbon dioxide in solution is in direct proportion to the partial pressure of the carbon dioxide.

It is easily seen that the open leg of the manometer is subject to variations in atmospheric pressure. This is corrected for by the use of a thermobarometer containing only water. The manometer changes on the thermobarometer are used to correct the samples for temperature or pressure variations.

Flask Calibration. The flasks and manometers are usually calibrated as units. The flask and manometer calibration is designed to yield the gas volume to a fixed reference point on the manometer. In all the author's work with the Warburg, he has used a reference point of 150. The gas volume can be determined with mercury by the technique of Umbreit in "Manometric Techniques" or by the water method. The water method offers much to be desired in simplicity while sacrificing little in the way of accuracy. With care the flask-manometer volume can be determined to within 2 per cent.

Water Method.

1. Add 10 ml of water to the 125-ml flask or 1 ml to the 25-ml flask (V_1).

2. Allow the flask to come to temperature by shaking for approximately 15 min.

3. Raise the manometer fluid to the 300-μl index with both legs open for the 125-ml flasks and to the 250-μl index for the 25-ml flask (R_1).

4. Close the flask, and lower the manometer fluid in the closed leg to the 150-μl index (I).

5. Record the reading of the open leg (R_2).

6. Repeat two times to confirm the reading.

7. Add 100 ml of water to the 125-ml flask and 10 ml to the 25-ml flask (V_2).

8. Allow the water to come to temperature equilibrium by shaking for 15 min.

9. Raise the index with both manometer legs open to 300 μl for the 125-ml flask and to 250 μl for the 25-ml flask (R_3).

10. Close the flask, and lower the manometer fluid to the 150-μl index (I).

11. Record the reading on the open leg of the manometer (R_4).

12. Repeat two times to confirm the reading.

The data reflect the pressure change in the system with a fixed volume change of fluid. From this it is possible to determine the manometer-flask volume by the following equation:

$$V = \frac{V_2(I - R_4)}{R_2 - R_4} + V_1$$

Warburg Procedure. In the determination of oxygen uptake it is necessary to add 10 per cent KOH to the center wells which have been well greased around the top to prevent KOH creep. Do not use a silicone grease for the center-well lubrication as it is too hard to clean. A chloroform-soluble grease is best. In the 125-ml flasks use 1.0 ml KOH, while 0.2 ml is used in the 25-ml flasks. The KOH can best be placed into the center well with a long needle hypodermic. If one drop of KOH should be out of the center well, the pH of the microbial solution would rise to toxic levels.

After the KOH is in the center wells, the sample to be tested is added. Normally 10 ml of the microbial solution plus 10 ml of the substrate are added to the large flasks or 1 ml of each to the small flasks. The flasks are connected to the manometers and placed on the Warburg apparatus with both legs of the manometer open. The flasks are allowed to shake for 5 min to allow the fluid to come to temperature equilibrium before closing off the flask. The index is set at 150 μl and all flasks closed. At the end of 15 min or some suitable time period the closed leg of the manometer is adjusted to the 150-μl index and the open leg is read. This procedure is repeated for all flasks. At regular intervals the readings are repeated. If it should become apparent that the pressure change would prevent a reading being made at the next time interval, the flask is opened and the index reset at 150 μl and then closed.

Cleaning the Flasks. As soon as the run is complete, the flasks are removed from the unit and rinsed several times with water and dried in a 103°C oven. The flasks are cooled and rinsed with chloroform to remove the grease from the center well and the manometer joint. The flasks are dried once again at 103°C, cooled, and cleaned with dichromate cleaning solution. The flasks are rinsed many times with water, dried at 103°C, and cooled in an inverted position. At this point the flasks are ready for reuse.

Calculating the Data. It is possible to collect a large quantity of Warburg data in a short time on an 18-unit Warburg. The calculation of the data can often require more time than the taking of the data. The first data of importance are the thermobarometer readings. A series of TB readings could look like this: 150, 148, 147, 148, 152, ... In the time period between the first two readings the decrease of 2 μl reflects an increase in atmospheric pressure. All the flask readings must be corrected for this 2-μl decrease. In the next time period the TB decrease only 1 μl, requiring a 1-μl subtraction from all flask readings. The next TB correction is a 1-μl increase rather than a decrease. This is followed by a 4-μl increase. If the fed sample readings had been 150, 140, 131, 124, 120 for the same time period, the calculations would be as follows. The first decrease is 10 μl, but TB decreased 2 μl. The actual decrease was $-10 - (-2) = -8$. This meant a decrease of 8 μl pressure in the flask or an O_2 uptake of 8 μl. The second period gave $-9 - (-1) = -8$. The third period was $-7 - (+1) = -8$ and the fourth period was $-4 - (+4) = -8$. It can be seen that the change in the TB is always algebraically subtracted from the flask changes.

The data are now recorded in microliters of O_2 uptake. Multiplication of this number by the K factor yields O_2 uptake in milligrams per liter.

Normally, a four-column set of data is used for each flask. The actual readings are recorded in column 1, the TB corrected change in column 2, the summation of the microliters of O_2 uptake in column 3, and the milligrams per liter of O_2 uptake in column 4 (column $3 \times K$ factor). In most Warburg work this latter column is plotted to show the pattern of oxygen uptake. The Warburg curves can yield a large quantity of valuable information, especially in waste treatment.

SUGGESTED REFERENCE

1. Umbreit, W. W., R. H. Burris, and J. F. Stauffer, "Manometric Techniques and Related Methods for the Study of Tissue Metabolism," Burgess Publishing Company, Minneapolis, Minn., 1948.

CHAPTER 10

Synthesis

It has been pointed out that the microorganisms process organic matter for the sole purpose of creating new cells. It is nature's way of maintaining its biological populations. We have already seen some of the chemical reactions which are required to synthesize cellular components and how the microorganisms obtain energy for synthesis.

Protoplasm

Protoplasm is not a single chemical compound which is the same for all microorganisms, but rather is a mixture of hundreds of different compounds. It can vary within a single species over wide ranges, depending upon the physical and chemical environment. But if all conditions are optimum, the various groups of microorganisms do form protoplasm with a relatively constant empirical formula for each group.

Bacteria—$C_5H_7O_2N$
Fungi—$C_{10}H_{17}O_6N$
Algae—$C_5H_8O_2N$
Protozoa—$C_7H_{14}O_3N$

Chemical Elements

The empirical formulation of protoplasm has shown that the major elements in protoplasm are carbon (C), hydrogen (H), oxygen (O), and nitrogen (N). If protoplasm is to be formed, the substrate must supply each element in the right quantity. While C, H, O, N are the major elements of protoplasm, there are many other elements required such as phosphorus (P), sulfur (S), sodium (Na), potassium (K), calcium (Ca), magnesium (Mg), iron (Fe), molybdenum (Mo), Cobalt (Co), manganese (Mn), zinc (Zn), and copper (Cu). Without all these elements, it would be impossible for the microorganism to form protoplasm and to carry out metabolic reactions at their optimum rate.

Examination of the dry protoplasm of bacteria indicates the following approximate concentrations of the major elements.

112

Element	Concentration, %
C	49
H	6.0
O	27
N	11
P	2.5
S	0.7
Na	0.7
K	0.5
Ca	0.7
Mg	0.5
Fe	0.1

The other elements are in trace quantities only.

Compounds Produced

The chemical elements in protoplasm combine to form definite compounds. The major group of compounds formed are the proteins which exist by themselves and in combination with nucleic acids, lipids, and carbohydrates. It has been estimated that bacteria are 60 per cent proteins, 15 per cent nucleic acids, 20 per cent carbohydrates, and 5 per cent lipids. Like all classifications this is a gross oversimplification of the compounds produced. Actually, much can be gained in knowing the chemical structures for protoplasmic fractions. If the bacteria possess the enzymes for synthesizing certain structures, they can use those same enzyme structures for degrading organic compounds of similar chemical structure.

Proteins. Proteins are made up of various amino acids joined together in peptide linkages. The major amino acids found in protoplasm are given in Fig. 10-1.

Nucleic Acids. The proteins synthesized from the amino acids are combined with nucleic acids to form nucleoproteins which make up the heart of any microbial cell. The nucleic acids are not simple structures in themselves and represent some interesting problems in synthesis. The components of nucleic acids include adenine, guanine, cytosine, uracil, thymine, ribose, and phosphoric acid. In nucleic acid there are two purine molecules and two pyrimidine molecules with four ribose or desoxy ribose molecules and four phosphoric acid molecules.

Lipids. The lipid formation in microorganisms is straightforward, with the production of both short- and long-chain fatty acids. The lipids do not pose any problem in synthesis, being primarily a polymerization of acetate. Aside from the lipids in the lipoprotein cytoplasmic membrane, lipids represent a storage mechanism for excess nonproteinaceous food.

Polysaccharide. The biological polysaccharides include all the hexose

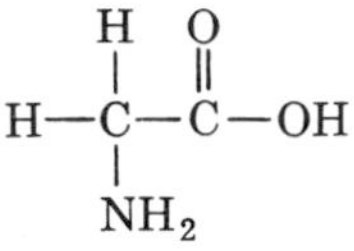

$$H-\underset{\underset{NH_2}{|}}{\overset{\overset{H}{|}}{C}}-\overset{\overset{O}{||}}{C}-OH$$

Glycine

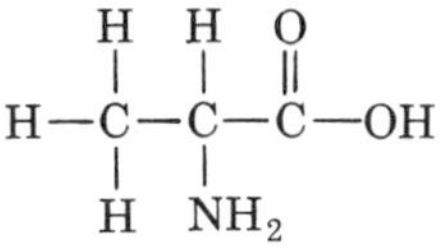

$$H-\underset{\underset{H}{|}}{\overset{\overset{H}{|}}{C}}-\underset{\underset{NH_2}{|}}{\overset{\overset{H}{|}}{C}}-\overset{\overset{O}{||}}{C}-OH$$

Alanine

$$H-\underset{\underset{HO}{|}}{\overset{\overset{H}{|}}{C}}-\underset{\underset{NH_2}{|}}{\overset{\overset{H}{|}}{C}}-\overset{\overset{O}{||}}{C}-OH$$

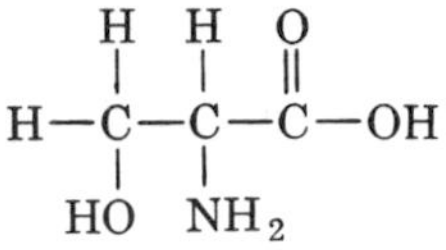

Serine

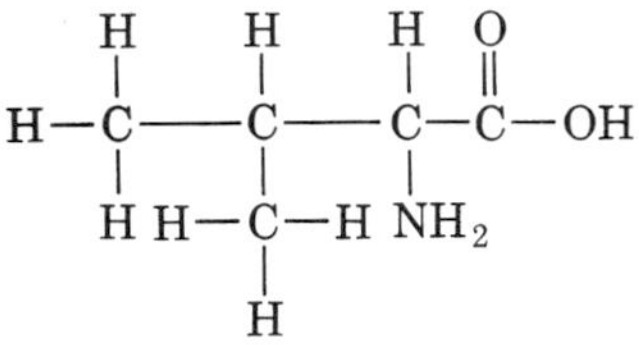

Valine

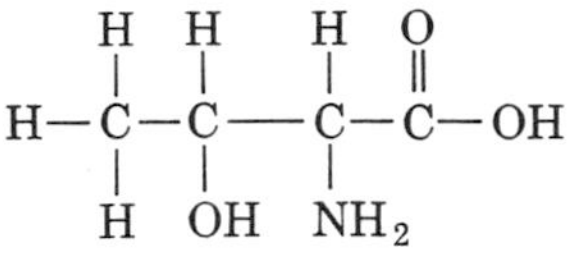

Threonine

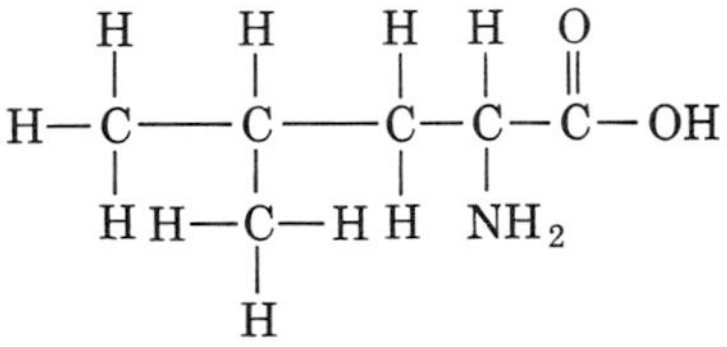

Leucine

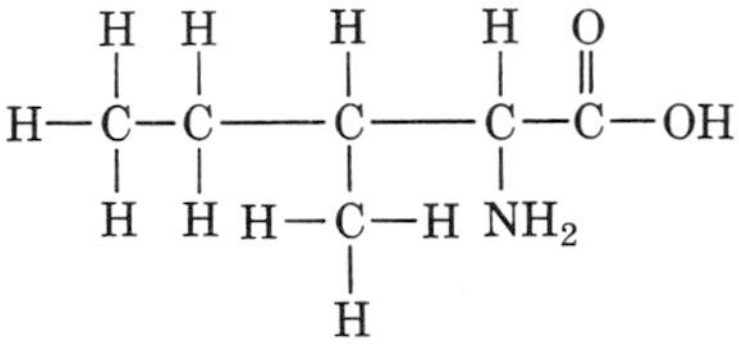

Isoleucine

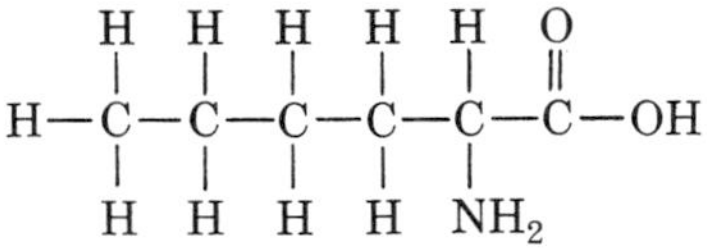

$$H-\overset{H}{\underset{H}{C}}-\overset{H}{\underset{H}{C}}-\overset{H}{\underset{H}{C}}-\overset{H}{\underset{H}{C}}-\overset{H}{\underset{NH_2}{C}}-\overset{O}{C}-OH$$

Norleucine

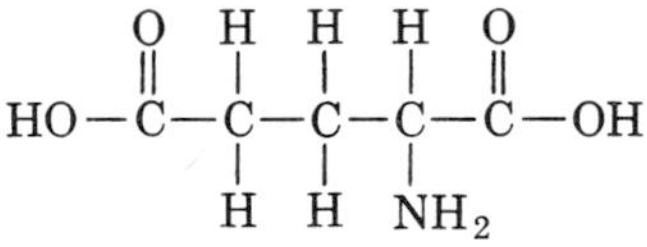

Glutamic acid

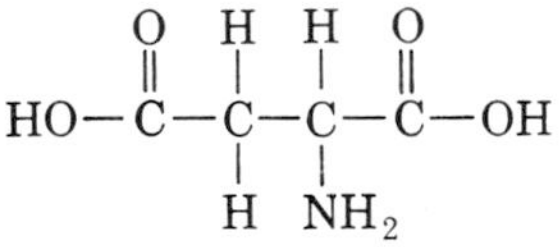

Aspartic acid

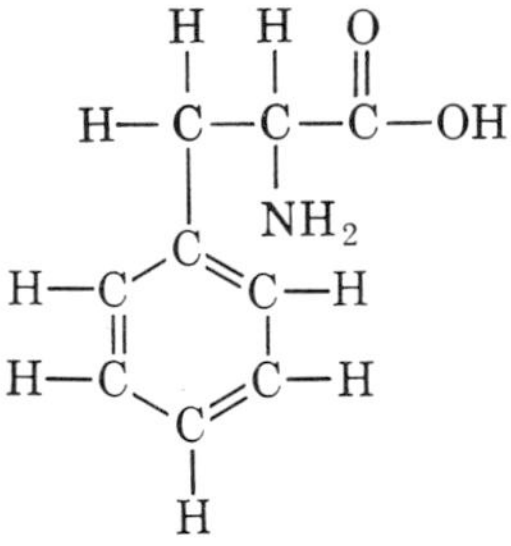

Phenylalanine

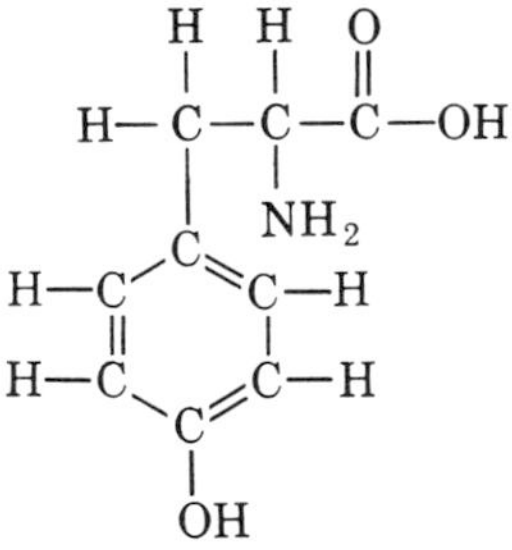

Tyrosine

Fig. 10-1. Common amino acids.

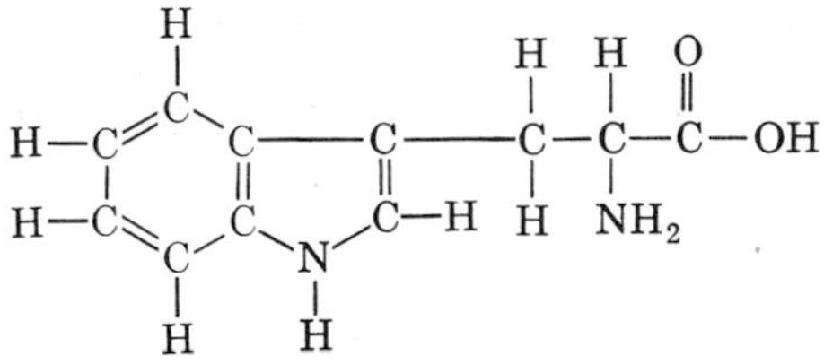

Tryptophan

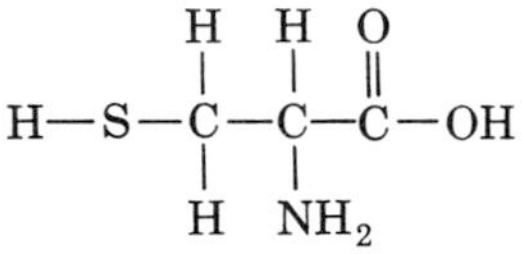

Cysteine

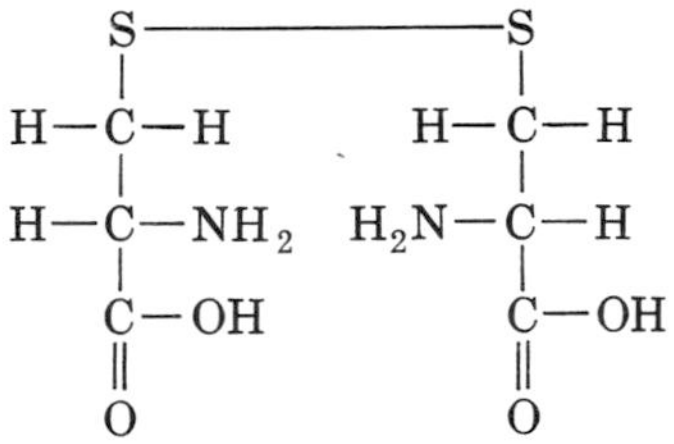

Cystine

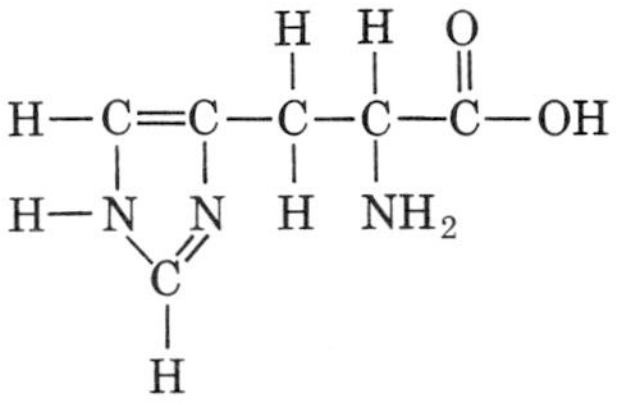

Histidine

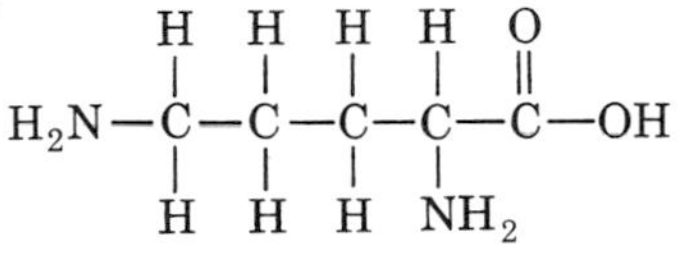

Lysine

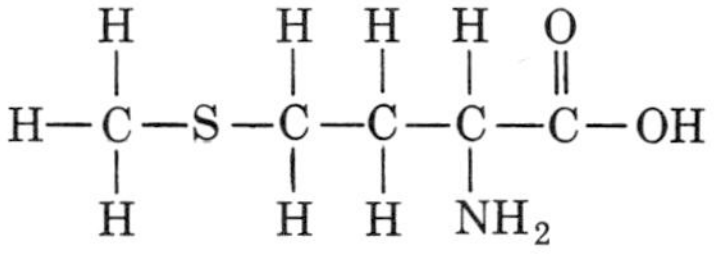

Methionine

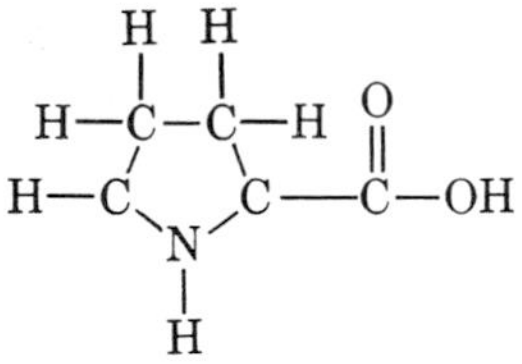

Proline

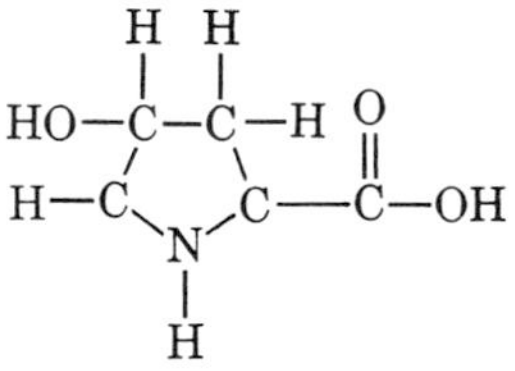

Hydroxyproline

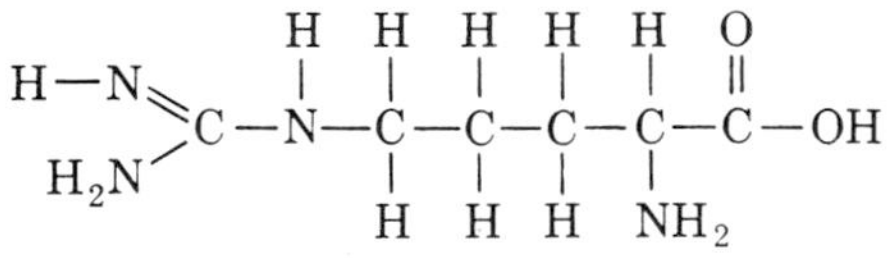

Arginine

FIG. 10-1. (*Continued*)

carbohydrates. The presence of pentoses has already been demonstrated in the nucleic acids. These structures pose no special problems in synthesis or in degradation.

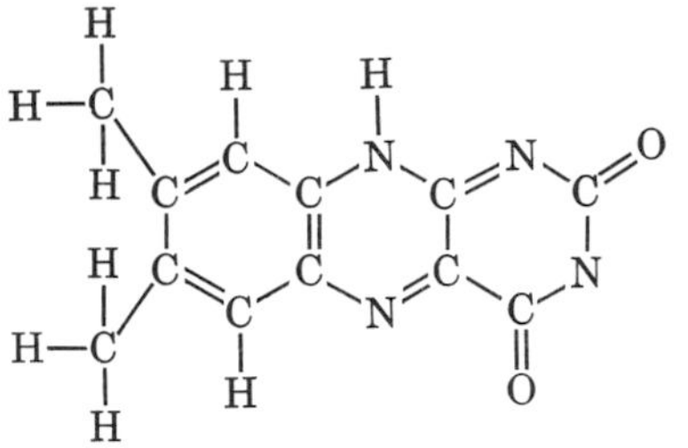

FIG. 10-2. Chemical structure of alloxazine.

Enzymes. The enzymes pose definite problems in synthesis. In DPN there is nicotinamide, adenine, ribose, and phosphate. In ATP, adenine, ribose, and phosphate make up the molecule. In FAD there appears a new structure, alloxazine (Fig. 10-2).

In CoA there are three additional compounds formed: thioethanolamine, beta alanine, and pantothenic acid (Fig. 10-3).

Para-aminobenzoic acid and pteridine are two components of folic acid which is involved in the synthesis of protoplasm (Fig. 10-4).

These structures are but a few of the compounds produced by the bacteria and other microorganisms. The common bacteria of sanitary significance can produce all these structures and others even more complicated from such simple materials as formic acid, ethanol, or glucose and inorganic salts.

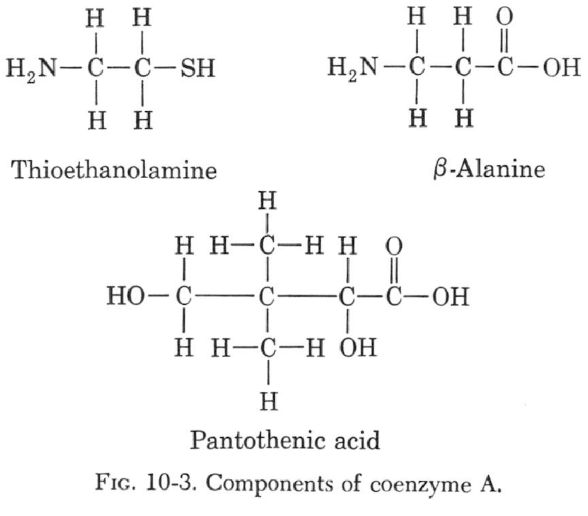

Thioethanolamine

β-Alanine

Pantothenic acid

FIG. 10-3. Components of coenzyme A.

Energy and Synthesis

Energy and synthesis are coupled reactions which cannot be separated. The maximum rate of energy expenditure occurs during maximum rate of synthesis. If synthesis should suddenly cease, the demand for energy would drop to a minimum. This means that since the organic matter in waste waters supplies both the energy and the building blocks for protoplasm, the maximum rate of removal of organic matter per unit of microorganisms occurs during maximum growth. The lowest rate of removal of organic matter per unit of microorganism occurs after all growth has ceased.

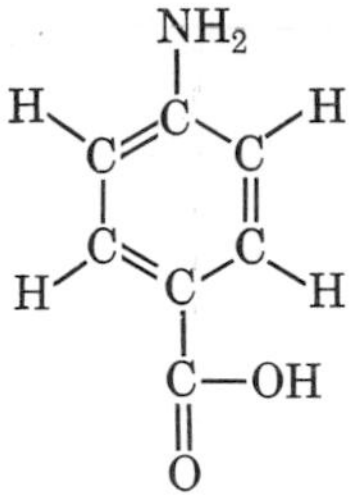

p-Aminobenzoic acid 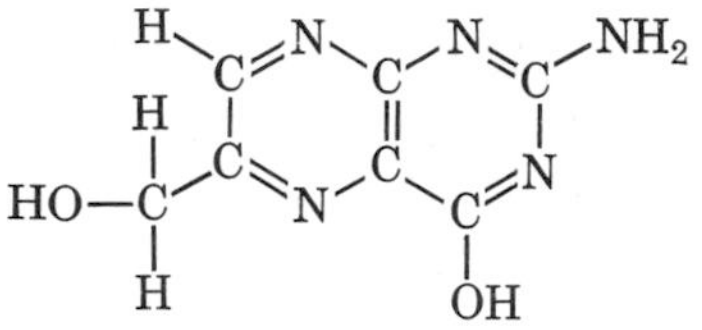   Pteridine

FIG. 10-4. Two compounds in folic acid.

Quantitatively speaking, the synthesis of a unit mass of protoplasm requires so many transformations that the energy requirements are the same regardless of the substrate being metabolized. Metabolism of a unit of energy in glucose will produce the same amount of protoplasm per unit of time as a unit of energy in phenol, ethanol, or any organic compound. This factor is of considerable importance in the design of biological waste treatment systems.

CHAPTER 11

Growth

One of the most important aspects of sanitary microbiology is the understanding and the control of microbial growth. The survival of pathogenic microorganisms, as well as the stabilization of wastes, is related to growth or the lack of growth.

Growth Patterns

The growth of microorganisms follows a definite pattern which has been studied most extensively with bacteria and protozoa which multiply by binary fission. Binary fission results in each cell dividing into two new cells of equal ability to metabolize. The net result is growth by twos. Although the pattern based on numbers of viable organisms has been studied primarily by microbiologists, the sanitary microbiologist has become more interested in the pattern based on mass of organisms.

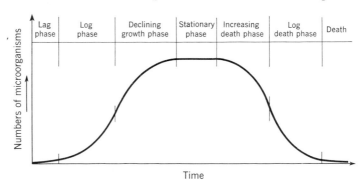

FIG. 11-1. Growth pattern based on numbers of microorganisms.

Number of Microorganisms. If a very small number of microorganisms are inoculated into a bacteriological culture medium, the growth pattern based on the number of viable organisms would look like Fig. 11-1. The growth pattern is normally divided into seven phases.

The initial phase is the *lag* phase. During this phase the microorganisms are adjusting to the medium and are not increasing in numbers.

The second phase is the *log* growth phase where growth is restricted only by the microorganisms' ability to process the substrate. The third phase is *declining* growth. The growth of the microorganisms in this phase is limited by the lack of food. The *stationary* phase is next. The microbial population remains level during this phase. *Increasing death* follows with the first decrease in population. *Log death* is next, with the rate of dying a function of the viable population. Finally, *death* takes over the culture and the growth cycle is complete.

Mass of Microorganisms. The growth pattern based on mass is shown in Fig. 11-2, and it has only three phases. Log growth starts as soon as the microorganisms come into contact with the substrate. The mass of the cells increases before numerical division occurs. Thus, the log growth phase based on mass covers both the lag and log phases based on numbers. The declining growth phase follows, with the endogenous phase

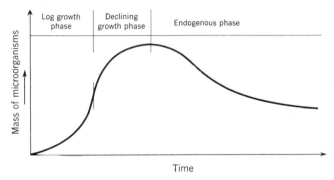

FIG. 11-2. Growth pattern based on mass of microorganisms.

last. With mass of cells there is no stationary phase. As soon as growth ceases, the cell mass starts to decrease.

Log Phase. In the log growth phase there is always an excess of food around the microorganisms. The rate of metabolism and growth is limited only by the microorganisms' ability to process the substrate. At the end of the log growth phase the microorganisms are growing at their maximum rate. At the same time they are removing organic matter from solution at their maximum rate.

Needless to say, the log growth phase has long intrigued the sanitary engineer, since the maximum rate of stabilization of organic matter occurs during the log phase. The use of the log growth phase for stabilizing wastes is limited by the fact that the organic concentration in the liquid surrounding the microorganisms must be high if the log growth phase is to be maintained. This means that it is impossible to produce a stable effluent as long as the microorganisms are in log growth. In aerobic systems the maximum rate of stabilization demands the maximum rate of oxygen. The low rate of oxygen transfer by diffused-air equipment limits the rate of growth in the log growth phase and has thwarted the sanitary engineer's use of log growth in waste stabilization.

Declining Growth. The limitation of food causes the rate of growth to decrease in the declining growth phase. As the microorganisms lower the food concentration, the rate of growth becomes less and less. Microbial growth in the declining phase is most often used for waste stabilization by sanitary engineers.

Endogenous Phase. When growth ceases, the food concentration is at a minimum. The small quantity of organic matter still in solution is in equilibrium with the microorganisms. As the microorganisms demand more food, they are forced to metabolize their own protoplasm, as well as slowly decreasing the food concentration in solution. The mass of microorganisms and the food concentration ratio remains constant during the endogenous phase. As the microbial mass decreases, the rate of metabolism decreases. Recently, the sanitary engineer has attempted to use the endogenous phase of metabolism for the complete stabilization of organic wastes.

Endogenous metabolism has proved of interest to the microbiologist. The prime question has been whether or not endogenous metabolism occurs constantly or only when food is insufficient for growth. This question has not been fully answered to date, but existing evidence with radiotracers has indicated that endogenous metabolism, respiration, occurs continuously but that during growth it is masked by new synthesis. From a practical point of view the endogenous metabolism reaction has little effect on the mass of protoplasm formed during the log growth phase but becomes significant during the declining growth phase.

Food-Microorganism Relationship

The total mass of microorganisms formed during the metabolism of varying concentrations of substrate over a fixed time period follows the pattern shown in Fig. 11-3.

Initially, the total mass produced is directly proportional to the substrate concentration. Eventually, the time period is not sufficient for complete metabolism and the mass produced per unit of time approaches a constant level. Most of the sanitary microbiologist's work occurs in the initial portion where metabolism is complete.

Aerobic-Anaerobic Growth

One of the least-used concepts in sanitary microbiology is the growth of microorganisms in aerobic conditions as opposed to anaerobic conditions. Aerobic metabolism of a substrate will yield 20 to 30 times the energy as anaerobic metabolism of the same substrate. This means simply that the growth of microorganisms per unit mass of organic matter will be 20 to 30 times greater under aerobic conditions than under anaerobic conditions. It also means that in order to support the same microbial population, 20 to 30 times the organic matter must be metabolized anaerobically rather than aerobically, the reason being the high-energy content of the end products of anaerobic metabolism. The utilization of the anaerobic bacteria's ability to metabolize organic matter at

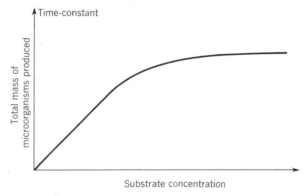

FIG. 11-3. Total mass of microorganisms produced for varying substrate concentration at constant time for growth.

a faster rate than aerobic bacteria has practical significance to the sanitary engineer.

Temperature

One of the most important factors affecting microbial growth is temperature. It has been observed that bacteria grow quite slowly at low temperatures but increase their rate of reaction as the temperature increases. It has been generally stated that the rate of microbial growth doubles with every 10°C increase in temperature up to the limiting temperature. The growth reactions are normal chemical reactions which follow definite patterns. The two patterns which are interposed with microorganisms are the increased rate of reaction with increased temperature and denaturation of specific proteins at definite temperatures. When these two phenomena are overlaid, we find that at low temperatures the denaturation reaction is insignificant. As the temperature approaches 35°C, the denaturation reaction becomes significant in most microorganisms. As the temperature is increased above 35°C, the denaturation reaction soon predominates and the microorganisms' rate of growth rapidly falls off to zero. The normal rate of growth of bacteria is illustrated in Fig. 11-4.

There are some microorganisms which can live above the temperature where most microorganisms die off. It has been found that these heat-tolerant microorganisms have proteins which resist denaturation at the lower temperatures. At 60 to 65°C the heat-resistant proteins are denatured and even the heat-tolerant microorganisms soon die off. The microorganisms which grow best at the elevated temperature range between 55 and 65°C are called *thermophilic* microorganisms. Needless to say, their rate of metabolism is very high at these temperatures. The majority of microorganisms which grow best at the lower temperature are called *mesophilic* microorganisms. The optimum temperature for the mesophilic bacteria is around 35°C; they die at 40 to 45°C.

Most microorganisms cannot grow in low temperatures since the water

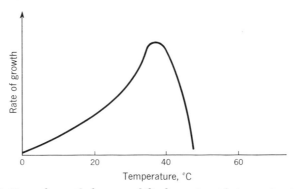

FIG. 11-4. Rate of growth for mesophilic bacteria with increasing temperature.

which makes up 80 per cent of the cell freezes and prevents further reaction. Some few microorganisms with a minimum of water have the ability of withstanding temperatures slightly below freezing and are known as *psychrophilic* microorganisms. The rate of growth and metabolic reactions of the psychrophilic microorganisms are very slow.

Culture Media

One of the most important aspects of microbial growth is the isolation and the quantitative enumeration of the microorganisms present in biological fluids. Two basic types of culture media are used, liquid and solid. The solid media are preferred for isolation and counting, but there have been instances where liquid media were required.

There are also two types of media within the liquid-solid classification, conventional dehydrated and synthetic. The conventional dehydrated media are the most common media used in microbiological studies. In sanitary engineering, the use of conventional dehydrated media has reached the stage of a "sacred cow." In many instances synthetic media have tremendous advantages over the conventional media. As sanitary microbiology grows up, there will be greater use and greater acceptance of synthetic media.

Liquid Media. Liquid media are used primarily as a general culturing media and for biochemical tests in identification. They are also used for counting by the most probable number (MPN) technique. Unfortunately, the number of tubes required for an accurate MPN is so great that this technique has limited value.

Liquid media have been used to stimulate growth of specific bacteria prior to isolation. This was especially valuable in research on floc-producing bacteria in activated sludge. Many of the complex protozoa can be grown only on liquid media. The same is true of some algae. Thus it is that liquid media have their value.

Solid Media. The most important aspect of solid media is their retention of growth at a single spot. This permits the ready isolation of a single species of bacteria or fungi. Since the solid media prevent the microbes from wandering far, all growth goes into a single spot. Soon sufficient growth has occurred so that it is possible to see the colony which has resulted from a single organism.

Dehydrated Media. The production of dehydrated culture media has done much to standardize the bacteriological culturing techniques. It is possible to buy standardized media from either Difco or BBL time and time again and be ensured of a homogenous media with uniform characteristics. The most common dehydrated media of interest to sanitary microbiologists are tryptone glucose agar, eosin methylene blue agar, lactose broth, and nutrient broth. It is possible to prepare these media merely by adding water and sterilizing. Full directions are printed on each bottle so that the media can always be prepared correctly.

Synthetic Media. Synthetic media are becoming of more value in sanitary microbiology. In industrial waste treatment bacteria must grow entirely on the organic fraction plus inorganic salts. Often it is desirable to isolate the predominant bacteria or fungi. This is accomplished by adding the waste or its major component to a nutrient salt solution and 1.5 per cent agar for a solidification agent. The microorganisms are isolated by either streak plates or spot plates. Normally in synthetic media the organic concentration is set at 5.0 gm/liter. The nutrient salts in most areas can consist of ammonium phosphate plus tap water. The ammonium phosphate concentration used is usually around 1.0 gm/liter. If it is desired to synthesize the medium completely, the following salts can be used in 1 liter of distilled water:

Salt	Grams
$FeSO_4 \cdot 7H_2O$	0.1
$MnSO_4 \cdot 2H_2O$	0.1
$NaCl$	2.0
$MgSO_4 \cdot 7H_2O$	0.2
$CaCl_2$	1.0
KH_2PO_4	1.0
K_2HPO_4	1.0
$(NH_4)_2HPO_4$	1.0
Na_2MoO	0.1

The pH must be adjusted to the proper range after sterilization. Normally the pH can be controlled by adjusting the two potassium phosphates. Increasing monobasic phosphate lowers the pH, while dibasic phosphate causes a rise in pH.

Inhibitory Media

It is possible to add inhibiting chemicals such as dyes, antibiotics, or metallic salts to prevent growth of extraneous microorganisms. The most common inhibitory media in sanitary microbiology are eosin methylene blue agar (EMB) and brilliant green bile broth (BGB). The two dyes, eosin and methylene blue, inhibit the growth of most bacteria on EMB agar. The presence of lactose stimulates the coliforms to overcome the toxic effect of the dyes. In BGB broth the brilliant green dye inhibits noncoliforms while the bile depresses surface tension and prevents growth of aerobic spore formers which can also metabolize lactose.

Fungi Media

Under normal biological growth conditions bacteria will always overgrow fungi. This makes it very difficult to isolate fungi from bacteria-fungi mixtures. Fungi are best isolated by adjusting the media to stimulate the fungi while inhibiting the bacteria. The fungi prefer carbohydrates to proteins so that glucose is usually present in the media. Fungi can grow easier at low pH than bacteria, so the media pH is usually adjusted to 4.5. Growth of fungi at normal pH is encouraged by the use of antibiotics to inhibit bacterial growth. Media for growth of fungi are readily available in dehydrated form.

Algae Media

Although algae grow in completely inorganic media, they have been grown only slightly in pure culture. Algae are grown best in mixed cultures in liquid media in which carbon dioxide is slowly bubbled. A media rich in inorganic nitrogen and phosphate will satisfy for good algae growth.

Protozoa Media

Protozoa are almost as difficult to grow in pure culture as algae. The complexity of protozoa requires a liquid media with ample oxygen and a heavy concentration of organic matter. Most of the media have been synthesized from individual amino acids. Experience has indicated that the common protozoa can be grown in normal dehydrated media such as nutrient broth or tryptone glucose broth. The growth is not as luxuriant as in some of the special media, but the dehydrated media are easier to prepare.

Isolation of Microorganisms

Bacteria. Bacteria are isolated primarily on agar media, using either streak plates or spot plates. In both methods the agar is solidified prior to use. Streak plates are made by taking a loop needle with the inoculum and making three short streaks across the surface of the agar on one edge of the plate. The needle is sterilized, cooled, and the ends of the previous streaks are crossed with three new streaks. The procedure is repeated two more times. The streak pattern is shown in Fig. 11-5. The streaks and restreaking result in diluting

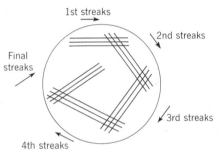

FIG. 11-5. A pattern for streak plates.

the microorganisms along the streaks. It is possible to find individual colonies along the secondary streaks for isolation.

One of the newest techniques for isolating pure bacteria cultures is the spot-plate technique. The spot-plate method actually allows quantitative enumeration, as well as individual isolation. The agar plates are dried at room temperature for 48 hr prior to use. Serial dilutions are made of the sample to yield a dilution containing approximately 2,000 bacteria per ml. Five 0.02-ml spots are pipetted on one-half of an agar plate with the aid of a microscrew attachment. The spots are allowed to be absorbed by the agar before being placed in the incubator. It is possible to set up duplicate counts or two dilutions on a single plate. The microorganisms grow on the surface of the agar as individual colonies (Fig. 11-6). This permits isolation of all predominant bacteria if desired.

At first glance it would appear that the use of 0.02-ml spots would be hard to measure accurately. The use of the microscrew attachment

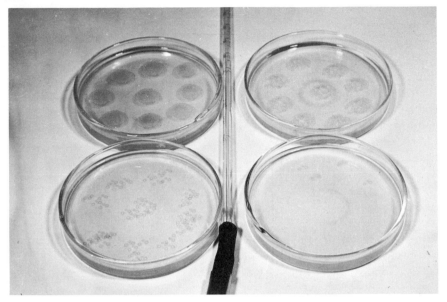

FIG. 11-6. Spot-plate cultures at four serial dilutions with 0.2-ml pipette and screw attachment.

makes this job very easy and quite accurate. A detailed study on this technique has shown that the spot-plate technique can yield more accurate results than the standard method pour plate. The novice bacteriologist can master the spot-plate technique more rapidly than the pour-plate techniques. For this reason the author uses the spot-plate technique for all routine bacteriological counting in samples containing

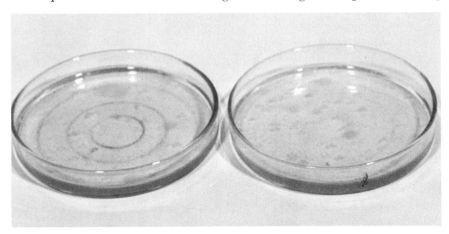

FIG. 11-7. Pour-plate bacteria colonies.

over 1,000 bacteria per ml. It is obvious that this technique is not satisfactory if the counts are under 1,000 per ml.

Pour-plate Counts. The standard method for enumerating bacteria is the pour-plate counting method (Fig. 11-7). In this method 1 ml of a diluted sample containing between 30 and 300 organisms per milliliter is added to a sterile petri dish. Approximately 10 to 15 ml of sterile media such as tryptone glucose agar in a molten state at 45°C is poured over the sample in the petri dish. The sample and the agar are thoroughly mixed and the agar allowed to solidify. The plates are incubated 24 to

 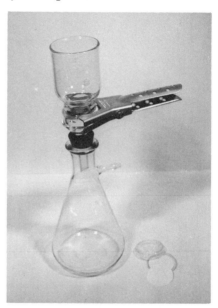

FIG. 11-8. Quebec colony counter for enumeration of bacteria on agar plates. FIG. 11-9. Membrane filter equipment.

48 hr and then counts are made using a Quebec colony counter (shown in Fig. 11-8).

The major problem in pour plates is in distribution of the sample uniformly throughout the agar. Considerable practice is required to get this technique perfected. The second problem is differentiating particulate matter in the agar from bacteria colonies. The third problem lies in the human error of having to count all of the bacteria on the entire plate. Even with a Quebec colony counter and a count recorder, it is difficult to make all counts accurately.

Membrane-filter Counts. One of the newest counting methods is the use of the membrane filter (Fig. 11-9). In this method the bacteria are filtered from solution onto the surface of a cellulose acetate filter. The filter is then placed on an absorbent pad containing the necessary nutrients in a special plastic petri dish. The plate is incubated for 24 hr and then counted. The membrane filters usually have a grid printed on them for ease of counting.

The membrane filter has opened up entirely new areas of quantitative microbiology, but it is far from a complete answer. Considerable controversy has raged in the field of sanitary microbiology over the membrane filter and its use for coliform analyses. Anyone who has ever run a membrane-filter count knows it is not a simple or an easy technique, but rather a complex one requiring special equipment and a high operating cost. Most sanitary microbiologists who run only a few bacteria counts a day do not like the membrane-filter technique, while those who run a large number of analyses would not have anything else. The reason for this lies in the preparation of media and petri dishes. In an effort to simplify the procedures and make them more attractive to the small user, the manufacturer has made kits which contain a sterilized petri dish, a sterilized nutrient pad, and sterilized media (Fig. 11-10). These kits take all the hard work from the membrane-filter method. Once the test has been run, the petri dish and filter are discarded. The cost of the complete kits is high by comparison with other methods, but the time saved in preparation and cleaning up justifies the additional expense. As more prepared media and simplified techniques become available, the membrane filter will be used more and more.

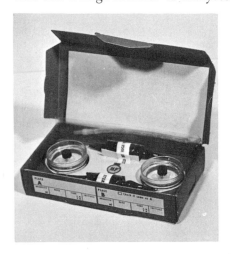

FIG. 11-10. Presterilized membrane-filter kit for field use.

The membrane filter works best on a very dilute sample and is designed to pass a relatively large volume of water through the filter to give uniform distribution. The filter holder and the filter funnel must be sterile. Alcohol sterilization is commonly used between samples. The sterile filter is aseptically placed on the filter holder with sterile forceps. The sample is filtered under vacuum and the filter carefully removed with sterile forceps. The filter is rolled onto the nutrient pad containing approximately 2 ml of liquid media. The media are pulled into the filter by capillary action. The bacteria on the surface grow as discrete colonies and can be easily removed for isolation and identification.

Most Probable Number (MPN). One of the crudest quantitative methods for enumerating bacteria is the most probable number (MPN). It is a statistical count based on the probable number of bacteria growing in several tubes in a series of serial dilutions. The MPN technique has been used in sanitary microbiology for coliform enumerations. In coliform analysis it has been the practice to use the five-tube, three-dilution method. The ability of the coliforms to ferment lactose is used as the indicator for presence of coliforms. In the lactose broth tubes an inverted vial is inserted to trap the gas resulting from the lactose fermentation. One of the most baffling experiences to the novice is the first time he sterilizes media with fermentation tubes. As he removes the tubes from the autoclave, he notices that the inverted vial is full of gas already and immediately jumps to the conclusion that he did something wrong. There is always a definite sigh of relief when the tubes cool and the fermentation vials are filled.

Five tubes of lactose broth are inoculated for each of three serial dilutions and incubated for 24 hr at 35°C. Often it is necessary to set up four or five serial dilutions to get three serial dilutions with valid results. The production of gas is considered presumptive evidence of coliforms. The MPN value is calculated from Table 11-1. For high MPN values, set the middle values in the middle column, and read the MPN value and multiply it by the dilution factor of the middle tube. A set of data in which a 10^5 dilution gave five positives, 10^6 gave four positives, and 10^7 gave zero positives would have an MPN reading of 5 to 4 to 0, indicating a value of 130. This value is multiplied by 10^6 to give the true MPN, 130×10^6 coliforms/100 ml.

The MPN value is not an absolute number, but it has been so abused by sanitary engineers that the MPN value has almost taken absolute significance. A single MPN value suddenly becomes an absolute number. The most apparent symptom of the absoluteness of the MPN value was in the evaluation study of the membrane-filter technique for coliform analyses. Some microbiologists felt the membrane-filter technique was no good because it failed to reproduce the MPN value. Actually, the membrane filter is the more accurate of the two methods. The comparison of a more accurate method to a less accurate method and then criticizing the more accurate method for failing to agree with the less accurate method is the perfect example of illogical logic that periodically crops up in scientific research.

The MPN method has been used for a large number of years. Although the method is not precise, it has yielded results which are satisfactory

TABLE 11-1. MOST PROBABLE NUMBERS (MPN) OF COLIFORM BACTERIA
PER 100-ML SAMPLE
(Tubes Showing Gas of 5 Tubes with 3 Serial Dilutions,
10-, 1-, and 0.1-ml Samples)

10	1	0.1	MPN	10	1	0.1	MPN	10	1	0.1	MPN	10	1	0.1	MPN	10	1	0.1	MPN	10	1	0.1	MPN
0	0	0	0	1	0	0	2	2	0	0	4.5	3	0	0	7.8	4	0	0	13	5	0	0	23
0	0	1	1.8	1	0	1	4	2	0	1	6.8	3	0	1	11	4	0	1	17	5	0	1	31
0	0	2	3.6	1	0	2	6	2	0	2	9.1	3	0	2	13	4	0	2	21	5	0	2	43
0	0	3	5.4	1	0	3	8	2	0	3	12	3	0	3	16	4	0	3	25	5		3	8
0	0	4	7.2	1	0	4	10	2	0	4	14	3	0	4	20	4	0	4	30	5	0	4	76
0	0	5	9	1	0	5	12	2	0	5	16	3	0	5	23	4	0	5	36	5	0	5	95
0	1	0	1.8	1	1	0	4	2	1	0	6.8	3	1	0	11	4	1	0	7	5	1	0	33
0	1	1	3.6	1	1	1	6.1	2	1	1	9.2	3	1	1	14	4	1	1	21	5	1	1	46
0	1	2	5.5	1	1	2	8.1	2	1	2	12	3	1	2	17	4	1	2	26	5	1	2	64
0	1	3	7.3	1	1	3	10	2	1	3	14	3	1	3	20	4	1	3	31	5	1	3	84
0	1	4	9.1	1	1	4	12	2	1	4	17	3	1	4	23	4	1	4	36	5	1	4	110
0	1	5	11	1	1	5	14	2	1	5	19	3	1	5	27	4	1	5	42	5	1	5	130
0	2	0	3.7	1	2	0	6.1	2	2	0	9.3	3	2	0	14	4	2	0	22	5	2	0	49
0	2	1	5.5	1	2	1	8.2	2	2	1	12	3	2	1	17	4	2	1	26	5	2	1	70
0	2	2	7.4	1	2	2	10	2	2	2	14	3	2	2	20	4	2	2	32	5	2	2	95
0	2	3	9.2	1	2	3	12	2	2	3	17	3	2	3	24	4	2	3	38	5	2	3	120
0	2	4	11	1	2	4	15	2	2	4	19	3	2	4	27	4	2	4	44	5	2	4	150
0	2	5	13	1	2	5	17	2	2	5	22	3	2	5	31	4	2	5	50	5	2	5	180
0	3	0	5.6	1	3	0	8.3	2	3	0	12	3	3	0	17	4	3	0	27	5	3	0	79
0	3	1	7.4	1	3	1	10	2	3	1	14	3	3	1	21	4	3	1	33	5	3	1	110
0	3	2	9.3	1	3	2	13	2	3	2	17	3	3	2	24	4	3	2	39	5	3	2	140
0	3	3	11	1	3	3	15	2	3	3	20	3	3	3	28	4	3	3	45	5	3	3	180
0	3	4	13	1	3	4	17	2	3	4	22	3	3	4	31	4	3	4	52	5	3	4	210
0	3	5	15	1	3	5	19	2	3	5	25	3	3	5	35	4	3	5	59	5	3	5	250
0	4	0	7.5	1	4	0	11	2	4	0	15	3	4	0	21	4	4	0	34	5	4	0	130
0	4	1	9.4	1	4	1	13	2	4	1	17	3	4	1	24	4	4	1	40	5	4	1	170
0	4	2	11	1	4	2	15	2	4	2	20	3	4	2	28	4	4	2	47	5	4	2	220
0	4	3	13	1	4	3	17	2	4	3	23	3	4	3	32	4	4	3	54	5	4	3	280
0	4	4	15	1	4	4	19	2	4	4	25	3	4	4	36	4	4	4	62	5	4	4	350
0	4	5	17	1	4	5	22	2	4	5	28	3	4	5	40	4	4	5	69	5	4	5	430
0	5	0	9.4	1	5	0	13	2	5	0	17	3	5	0	25	4	5	0	41	5	5	0	240
0	5	1	11	1	5	1	15	2	5	1	20	3	5	1	29	4	5	1	48	5	5	1	350
0	5	2	13	1	5	2	17	2	5	2	23	3	5	2	32	4	5	2	56	5	5	2	540
0	5	3	15	1	5	3	19	2	5	3	26	3	5	3	37	4	5	3	64	5	5	3	920
0	5	4	17	1	5	4	22	2	5	4	29	3	5	4	41	4		4	72	5	5	4	1600
0	5	5	19	1	5	5	24	2	5	5	32	3	5	5	45	4	5	5	81	5	5	5	

SOURCE: "Standard Methods for the Examination of Water and Wastewater," 11th ed., American Public Health Association, New York, 1961.

for proper sanitary control. In spite of its inaccuracies and the large number of tubes required, the MPN method is not going to be immediately displaced for coliform analyses and has definite value in sanitary microbiology.

SUGGESTED REFERENCE

1. "Standard Methods for the Examination of Water and Wastewater," 11th ed., American Public Health Association, New York, 1961.

CHAPTER 12

Death

The sanitary microbiologist is as interested in the death of microorganisms as he is in their growth. The prime function of the sanitary microbiologist is the prevention of the spread of disease through water, milk, or food on a mass scale. This can be done best by killing the microorganisms before they reach the individual.

Pattern of Death

We have already seen the normal pattern of death in the growth cycle. The pattern of death will follow the sigmoid curve if death is due to a single cause (Fig. 12-1). But if death should be due to several factors,

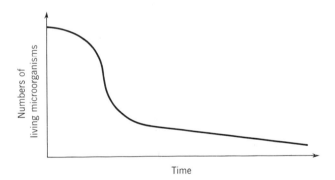

FIG. 12-1. Normal or autocatalytic death curve.

the pattern of death will not follow any special function or mathematical relationship. Normally with multiple death factors the death curve will be a multiple function.

Heat

One of the most widely used methods for destroying microorganisms is heat. Heat causes the protein fractions in the cell to undergo denaturation, with a loss of cellular organization and death. Heat can be applied as dry heat or as moist heat as in steam. Most vegetative cells are easily killed by raising the temperature to 45°C. Only the thermophilic microorganisms can survive. Raising the temperature to 85°C will kill the vegetative thermophilic cells. Spores are the hardest form of biological life to kill by heat. The thick spore coating protects the spore from heat until the heat is very high for a prolonged period.

Normal bacteriological procedures for dry-heat sterilization require a temperature of 170°C for a period of 1 hr. Dry-heat sterilization is used for glassware and metal objects not containing liquids. The high-temperature and long time period is required for the heat to penetrate all areas of the object and around the object. Steam sterilization is used for liquids and is carried out at 121°C, 15 psi pressure for 15 min. The elevated pressure prevents loss of the liquid from the system. The lower temperature and shorter period required for steam sterilization than for dry-heat sterilization is due to the greater penetration of moist heat than dry heat.

Disinfection

Two of the most misused terms in sanitary microbiology are sterilization and disinfection. Sterilization is the complete destruction of all microbial life, while disinfection is the complete destruction of all pathogenic microorganisms. These two terms are similar but quite different. The most common misuse is by the chemical disinfectant manufacturers who often claim their product sterilizes when it merely disinfects.

Disinfection is usually brought about by chemical agents such as chlorine, phenol, cationic detergents, and the like. These materials can be used to sterilize a biological fluid, but it usually is impractical to do so merely from the material requirements. Disinfection is far more economical and practical.

Disinfection is effective because most of the pathogenic microorganisms are more sensitive microorganisms than the nonpathogens. This is true only in the case of the vegetative cells. The spores and cysts of the pathogens are usually more resistant than vegetative cells of nonpathogens. The rate of killing by a disinfectant is primarily a function of the disinfectant concentration and a characteristic of the microorganism, namely, the point at which the life function is destroyed, as well as temperature. Increasing the disinfectant concentration or the temperature results in increased toxicity by a power function rather than by a simple arithmetic function.

Oxidizing Agents

Chlorine and chlorine derivatives are the most common oxidizing agents used in sanitary engineering. In dilute solutions the toxic reaction appears to be related to the oxidation of the sulfhydryl radical ($-SH$)

in the key enzymes. As the concentration of oxidizing agent becomes strong, the proteins undergo denaturation at the amino group. Two other oxidizing agents used in sanitary analyses are permanganate and di-chromate. Their toxicity is due primarily to the oxidizing reaction, al-though chromium is a toxic element in its own right.

Surfactants

Synthetic detergents have mild disinfecting properties. The three ma-jor classes of syndets are anionic, nonionic, and cationic (Fig. 12-2).

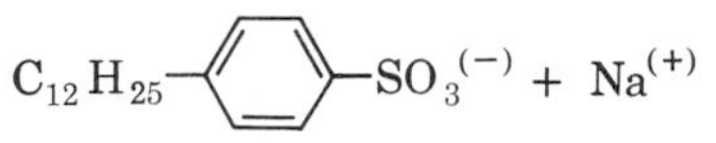

$$C_{12}H_{25}\text{—}\langle\!\!\!\bigcirc\!\!\!\rangle\text{—}SO_3^{(-)} + Na^{(+)}$$

Dodecyl benzene sulfonate

(*a*) Anionic

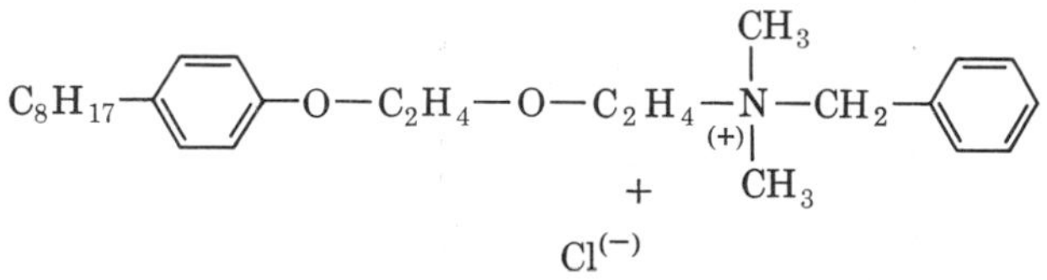

$$C_8H_{17}\text{—}\langle\!\!\!\bigcirc\!\!\!\rangle\text{—}O\text{—}C_2H_4\text{—}O\text{—}C_2H_4\overset{CH_3}{\underset{(+)|}{\text{—}N\text{—}}}CH_2\text{—}\langle\!\!\!\bigcirc\!\!\!\rangle$$
$$+ \qquad CH_3$$
$$Cl^{(-)}$$

p-tert-Octylphenoxyethoxyethyldimethyl ammonium chloride

(*b*) Cationic

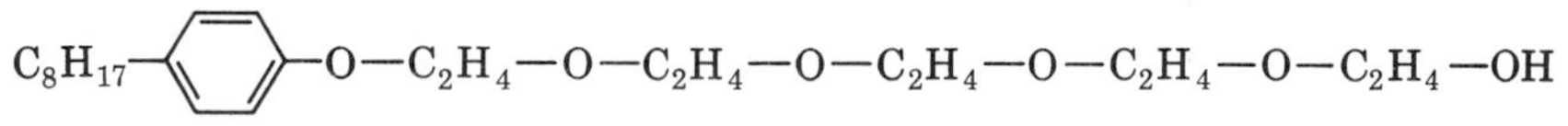

$$C_8H_{17}\text{—}\langle\!\!\!\bigcirc\!\!\!\rangle\text{—}O\text{—}C_2H_4\text{—}O\text{—}C_2H_4\text{—}O\text{—}C_2H_4\text{—}O\text{—}C_2H_4\text{—}OH$$

Octylphenoxy tetraethoxy ethanol

(*c*) Nonionic

FIG. 12-2. Three major classes of synthetic detergents.

The anionic have a negatively charged surfactant, while the nonionic has no charge and the cationic has a positive charge. In neutral solutions, microorganisms have a negative charge. The negatively charged micro-organisms attract the cationic syndets to their surface where the high-salt concentration results in damage of the cell membrane. The cationic and nonionic syndets are toxic only at very high concentrations due to destruction of the lipoprotein cytoplasmic membrane.

Heavy Metals

The heavy metals such as mercury, copper, silver, and arsenic are toxic to microorganisms because of their ability to tie up the proteins in the key enzyme systems. The heavy metallic salts prevent the proteins from reacting normally and even change the charge at certain points from negative to positive, causing repelling reactions rather than attracting reactions. If the concentration of the heavy metal is increased to quite a large amount, the surface of the cell becomes completely coated, preventing materials from entering the cell. Precipitation of the cellular protein may even occur.

Antimetabolites and Antibiotics

One of the first antimetabolites was sulfanilamide. This wonder drug has been followed by many, many more. Sulfanilamide was effective against those bacteria which could not synthesize their own para-amino-benzoic acid (PABA) needed for folic acid in the protein synthesis enzyme system. The concept of antimetabolites opened up an entirely new field of chemotherapy. The use of structurally related but nondegradable

$$p\text{-Aminobenzoic acid} \qquad \text{Sulfanilamide}$$

FIG. 12-3. Chemical structure of PABA and sulfanilamide.

compounds has proved most effective. The sulfonic acid radical was the key to the success of sulfanilamide. Its structural similarity to the carboxyl radical allowed it to react with the enzyme system normally reacting with the carboxyl radical (Fig. 12-3). The difference in structural configuration between the two radicals prevented completion of the reaction with the result that the enzyme system remained tied up with sulfanilamide and synthesis stopped.

At first, all microorganisms were affected by sulfanilamide, and it was tagged a "wonder" drug. Soon it became apparent that any microorganism capable of producing PABA on its own could produce enough PABA to displace the sulfanilamide by the old law of mass action and proceed with synthesis. Later, after sulfanilamide had been around long enough, some microorganisms adapted to using it as their substrate and could remove it from solution. As a result sulfanilamide lost its punch.

About this time the fungi-produced antibiotics made their debut. Penicillin was extremely effective against many pathogenic bacteria. It was found that penicillin prevented the passage of glutamic acid into the cell. Those bacteria which could synthesize their own glutamic acid were unaffected by penicillin. Like sulfanilamide, the presence of penicillin in large quantities stimulated the growth of bacteria which could metabolize penicillin.

Streptomycin was even more effective than penicillin. It reacted best under basic pH by blocking the entry of pyruvate into the citric acid cycle. There are other antibiotics of varying degrees of effectiveness. The most important aspect of antibiotics is that the presence of large quantities of antibiotics in use stimulates bacteria which not only are resistant but can degrade the antibiotic as a source of food.

pH

One of the best controls on microbial growth is pH. At low pH the hydrogen-ion concentration causes denaturation of the key enzyme proteins. Most microorganisms cannot survive below pH 4.0, but a few sulfate oxidizing bacteria can exist at a pH of 1.0. The same is true of the hydroxyl-ion concentration. As the pH rises over 9.5, the hydroxyl ion begins to exert a toxic effect. Few, if any, microorganisms can survive above pH 11. Control of pH at either a high or a low range can be used to prevent decomposition of stored waste matter until desired. Actually, pH control is the most significant economic control the sanitary microbiologist has over the growth and death of microorganisms.

CHAPTER 13

Population Dynamics

In nature microorganisms do not exist as pure cultures, but rather as mixtures. Each microorganism must compete with its neighbor in order to survive. The pattern of competition and survival in mixtures of microorganisms is set within certain limits. Since all of the sanitary microbiological phenomena deal with control of mixed microbiological populations, it is essential that the sanitary microbiologist have a good grasp of population dynamics.

Competition for Food

The prime factor in population dynamics is the competition for food. In order to grow, a microorganism must be able to derive a certain quantity of nutrients from the system in question. Thus, any mixed population of microorganisms contains only those microorganisms which have been successful in this competition. Two types of competition for food exist, competition for the same food and use of one organism by another organism for food.

Same Food. The most common competition is that for the same food. The microorganisms' ability to compete successfully for food under a fixed set of environmental conditions is a function of the metabolic characteristics of the microorganisms. The microorganisms which can process the maximum quantity of food at the maximum rate will predominate.

If two species of bacteria are placed in a nutrient solution which both can utilize, they will both grow. If one of the bacteria species cannot metabolize the nutrients completely, that species will not be able to survive to the same extent because of its lack of ability to obtain energy. In sanitary microbiology those bacteria which only metabolize organic matter partially give way to those with complete metabolic processes. In aerobic systems with a proper balance of nutrients this means that the bacteria which oxidize organic matter completely to carbon dioxide and water will survive and predominate over bacteria with incomplete metabolic patterns.

If two species of bacteria have the ability to metabolize the same food

136

at the same rate, they will survive together in equal *masses*. If one bacterial species is larger than the other one, the number of each species surviving is a function of the mass of cellular material. If bacteria species A is twice the mass of species B, then the bacterial numbers will yield two of species B for every one of species A, but their survival is the same based on mass. Normally, the larger bacteria will not be able to process organic matter at the same rate as the smaller bacteria. This results from the reduced surface area of the larger bacteria which prevents as rapid an uptake of organic matter.

Most bacteria are the same size and survive primarily on their unusual metabolic reactions. The *Pseudomonas* can metabolize almost every type of organic matter and survive in almost every environment. Thus it is that the *Pseudomonas* must be regarded as the prime bacterial genera responsible for the degradation of organic matter of sanitary significance. The *Alcaligenes* and *Flavobacterium* are almost as important as *Pseudomonas* in that they metabolize primarily proteins. Wherever proteins are found, there will be *Alcaligenes* and *Flavobacterium*.

This same type of competition for food also occurs with fungi, algae, and protozoa. But what about competition between fungi, bacteria, and the holophytic protozoa? These organisms have the same metabolic habits in that they take in soluble food and process it by a similar pattern. These microorganisms differ from each other in size and rate of metabolism. The bacteria are the smallest and metabolize the fastest. Fungi are next, and the holophytic protozoa are last. In very strong organic solutions under aerobic conditions all microorganisms will grow, but the bacteria will predominate. In weak organic solutions the holophytic protozoa will be unable to do little more than survive. The fact that bacteria will always predominate over fungi and holophytic protozoa under optimum environmental conditions is very important to the sanitary microbiologist in waste treatment.

Prey-Predator Relationship. One of the major competitions for food is between plants and animals. The plants process soluble food, while the animals process solid food. Thus, the plants become the food for the animals. In a true sense the plants and the animals do not compete for the organic matter in the system, but their survival is related to it.

The plants process the organic matter into new cells, which stimulates the animals. As the animals metabolize the plants, the plants are able to metabolize the organic matter to a low level. In a sense the animals help adjust the food-microorganism ratio by reducing the microorganism concentration, thereby stimulating greater utilization of the food. When the organic matter has all been depleted, the plant population is reduced by the animals. But soon the plant population is decreased too low, and the animals begin to die off. The plant and animal populations decrease in proportion to the available food. In nature the animals are unable to eat all the plants so that both forms survive as long as any organic matter remains.

The relationship between the plants and the animals is the secret of success in biological waste treatment systems. The animal forms keep the excess bacterial populations down to allow the lowest food concentration in the system and assist in producing a clarified effluent. Treatment is never complete unless there is a proper balance between plants and animals. The most widely used waste treatment system, the river, shows the equilibrium between plants and animals to its greatest extent.

Nature of Organic Matter

The nature of the organic matter does much to select the types of microorganisms which predominate. Carbohydrates stimulate both bacteria and fungi. Of the bacteria there are many genera which can grow on carbohydrates. Organic acids, aldehydes, ketones, and alcohols stimulate *Pseudomonas, Micrococcus, Bacillus, Achromobacter,* and others. We have already indicated that proteins stimulate the *Alcaligenes* and *Flavobacterium.* An inorganic solution naturally stimulates the growth of algae, while both bacteria and algae stimulate the protozoa.

Environmental Conditions

We have already discussed the effect of the environment on each group of microorganisms. In mixtures the reactions are much the same. At neutral pH, 6.5 to 8.5, the bacteria predominate over fungi. Below pH 6.5 the fungi are able to compete with the bacteria more successfully. At pH 4.0 to 5.0 the fungi predominate almost to the complete exclusion of the bacteria.

The oxygen tension affects predomination since it controls the energy yield. In aerobic systems the bacteria, fungi, and protozoa all grow very readily. The metabolic reactions always yield carbon dioxide, water, and new cells. But under anaerobic conditions the picture changes; the fungi and the protozoa cannot grow, and only the bacteria survive. The common bacteria are unable to metabolize the organic matter to completion under anaerobic conditions. As a result special bacteria grow up to finish metabolism. The sulfate-reducing bacteria grow if sulfates are present; otherwise the methane bacteria grow. These special bacteria exist under fixed environmental conditions which can be controlled by the sanitary engineer. The fact that fungi cannot grow under anaerobic conditions but bacteria can offers a method for shifting fungi predomination to bacteria predomination.

Temperature levels also affect predomination. Increased temperatures stimulate enteric bacteria. Very high temperatures kill the normal micro-organisms and permit only the thermophilic organisms to survive.

Secondary Bacteria Predomination

In mixed bacterial populations the microorganisms which utilize the substrate will grow rapidly. When the substrate has been removed, the microorganisms die and lyse, releasing many cellular components to the surrounding solution. These cellular components are predominantly pro-teins. As a result the *Flavobacterium* and *Alcaligenes* are able to grow in secondary predominance. The primary and secondary predominance relationship is shown in Fig. 13-1. This concept is very important in the treatment of industrial wastes which require highly specialized primary microorganisms. Too long a period of contact can result in reduced

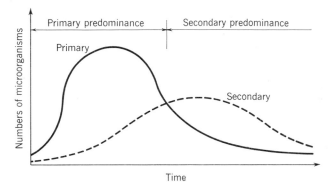

Fig. 13-1. Schematic diagram of primary and secondary bacteria predominance.

efficiency due to too low a population of primary microorganisms between feedings.

Protozoa Dynamics

The predomination of protozoa follows a fixed pattern more closely than the other groups of microorganisms. Because of the ease with which protozoa may be seen under the microscope, the protozoa are very valuable indicators of the over-all biological cycle.

The Mastigophora, flagellates, are never found in large numbers except in very freshly polluted waters. The phytoflagellates must compete with the bacteria for the soluble substrate and are never successful in this type of competition. The zooflagellates are more successful than the phytoflagellates because of their use of bacteria for food. But the zoo-flagellates are not as efficient as the free-swimming ciliates in obtaining bacteria and give way to the ciliates. As long as the bacterial population

is high, the free-swimming ciliates have a field day. With a decreasing
bacterial population, the free-swimming ciliates demand so much energy
that they give way to the stalked ciliates. The stalked ciliates attach themselves to solid particles and draw the food to them by rapidly moving
cilia. Their lower energy demand allows them to survive at very low
bacterial populations. Eventually, the system is so stable that the stalked
ciliates cannot obtain enough energy to survive. Rotifers and other higher
animals are the last microscopic organisms to survive. These microorganisms have the ability to utilize the nonsoluble fraction of the dead

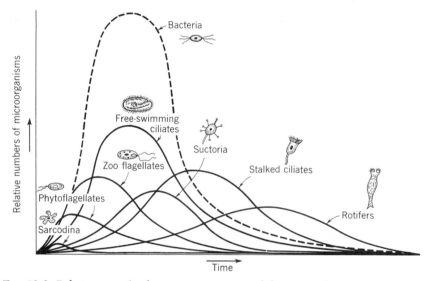

Fig. 13-2. Relative growth of microorganisms in stabilization of liquid organic wastes.
(*Note:* Numbers of microorganisms are not to same scale, but are expanded for clarity.)

bacteria, as well as other solid organic particles. The relative growth of
the protozoa is shown in Fig. 13-2.

Bacteria-Algae Relationship

One of the strangest relationships is between bacteria and algae.
These two microorganisms do not compete with each other for food, but
their activities are often dependent upon one another. Bacteria metabolize
organic matter under aerobic conditions to carbon dioxide and water.
Algae utilize carbon dioxide with the release of oxygen. The carbon
dioxide–oxygen relationship of the algae-bacteria symbiosis is used in
the treatment of waste waters in many sections of the country.

Close examination of the biochemical phenomena involved shows that
the bacteria metabolize complex organics in the presence of oxygen to

yield new cells, carbon dioxide, water, ammonia, and other inorganics (Fig. 13-3). The algae take the carbon dioxide, ammonia, and other in-

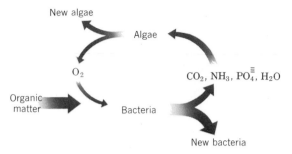

FIG. 13-3. Bacteria-algae symbiosis.

organics and convert them into new cells. Oxygen is released as a side reaction.

Special Cultures

One of the pet projects of many sanitary microbiologists has been the attempt to utilize special pure cultures to speed the rate of stabilization of certain wastes. Unfortunately, the concept of population dynamics shows the fallacies of this idea. Mixtures of microorganisms always produce a more stable result than single pure cultures.

Often a pure culture of a special bacteria is added to a mixed biological population to change the predomination. Actually, the predomination of microorganisms is the result of the environment. While it is possible to shift the population for a short period by the addition of a large mass of a pure culture, the predominance will reestablish itself in accordance with environmental conditions. Changes in the environment can produce predomination shifts without the addition of special organisms. Actually, the best source of microorganisms is soil. The soil can furnish all the microorganisms ever needed in waste disposal. My advice to all sanitary bacteriologists who seek a special culture is to look under their feet; the supply is inexhaustible.

CHAPTER 14

Pathogenicity

Some microorganisms are capable of producing diseases in man and in the plants and animals that man uses. These disease-producing microorganisms are known as *pathogenic*. The pathogenic microorganisms produce disease by upsetting the normal metabolic reactions in the host organisms. The most common form of pathogenicity is parasitism, where the pathogen derives its nutrients at the expense of the host organism. A second form of pathogenicity stems from the production of toxins by the pathogens.

Parasitism

Most microorganisms derive their nutrients from inert organic and inorganic matter, but there are some microorganisms which must have living tissue in order to grow. The microorganisms which utilize nonliving organics for their metabolic reactions are known as *saprophytes*. As will be shown later, the saprophytes are very important in sanitary microbiology for the stabilization of waste organic matter.

The parasitic pathogens depend upon the host organism for a proper environment for growth and reproduction. The parasites do not intentionally produce disease in the host organism since it eventually means death to the parasite either through adverse reactions in the host or through death of the host.

Parasitic Biochemistry

From a biochemical viewpoint the parasitic pathogens are actually less developed than the saprophytes. It may well be that the parasitic pathogens were actually saprophytes which did not have to synthesize all their necessary growth accessory substances for the normal synthetic reactions since the host organism supplied these materials. Growth over many generations within various hosts may have caused these microorganisms to lose eventually the power to produce the essential growth accessory substances and to become totally dependent upon the host to

142

supply the materials required for growth. It is the dependency of the parasitic pathogen upon the host that offers the potential method of controlling the pathogens.

Growth of the parasitic pathogen upon living organisms has made cultivation of the pathogens very difficult. It is necessary to build the cultivation media with all the necessary growth accessory substances in addition to the standard nutrients. In some instances it has been necessary to add fluids from living organisms or even living tissue in order to cultivate the desired microorganisms. The limited metabolic reactions of the pathogens as compared with the normal saprophytic microorganisms makes it difficult to stimulate growth of the pathogen and suppress the growth of the saprophyte in mixed biological systems. Highly specialized media have been devised in some instances where a characteristic reaction of the pathogen permits its growth over the saprophytes. Such media will not permit the quantitative enumeration of the pathogen since it normally inhibits the pathogen to a limited extent.

Viruses. The viruses are all parasitic pathogens with highly specific reactions. It has already been pointed out that the viruses are almost pure chemical identities, nucleic acids, which have the ability to reproduce only within a host. The viruses call upon the host to furnish all the necessary components for their cellular structure. The host is thus robbed of key materials needed for its cellular development. At the same time there is produced a foreign protein with a highly specific reaction. The specificity of the virus reactions makes it easy to identify which virus has infected the host. It is interesting to note the importance of environment in the growth of the viruses. A virus has the ability to remain virulent outside of the host even though it is not growing. It may pass through a series of secondary hosts with no reaction because the environment is not suitable. Once a proper environment is reached, the rate of growth of the viruses is very rapid. Viruses which infect plants will not infect animals, and vice versa, while some animals are affected by particular viruses and others are not. Thus it is essential to realize the chemical environment for growth of the different viruses.

The chemical nature of viruses makes them difficult to control with chemical agents. The only effective method of combating viruses has been through the production of antibodies in the host cells. For the most part antibodies are protein materials which react with the viruses and prevent their reproduction. The presence of the virus in the host stimulates the antibody production so that once the host has survived an attack by a particular virus it normally acquires immunity against future attacks by the same type of virus. Immunity normally consists in an automatic production of high concentrations of antibodies anytime a virus enters the host, resulting in immediate control of the virus before it can grow.

Rickettsiae. The rickettsiae are the intermediate stage between viruses and bacteria. They have been transmitted to man through the bites of ticks and mites. Like the viruses the rickettsiae require living tissue for growth and cultivation. But unlike the viruses they are readily susceptible to two of the major groups of antibiotics, chloramphenicol and the tetracyclines. The host animal has the ability to build up an immunity if it survives an infestation of rickettsiae. All rickettsiae are not pathogenic, but very little is known about the nonpathogenic rickettsiae.

Bacteria. The pathogenic bacteria have been studied in great detail because of their importance in the survival of man down through the ages. Only a few of the bacterial species are pathogenic, but more is known of these bacteria than is known about the saprophytic bacteria. Most of the pathogenic bacteria are transmitted by water, milk, food, or direct contact as a result of poor sanitation. The importance of controlling bacterial disease by proper treatment of water and sewage will be discussed in detail in later sections.

The treatment of bacteria-induced diseases is much simpler than that of viral diseases since the bacteria are more complex microorganisms and hence are more susceptible to damage. The sulfa drugs proved effective against many of the common pathogenic bacteria because the sulfa compound competed for para-aminobenzoic acid (PABA). Unfortunately, those bacteria which were capable of synthesizing PABA produced enough PABA to displace the sulfa compound from its site in the enzyme reaction and permitted growth to resume.

The advent of the antibiotics permitted chemical control of most of the pathogenic bacteria. Penicillin prevented the uptake of glutamic acid from the environment and hence proved effective against those bacteria which could not synthesize their own glutamic acid. Streptomycin affects the citric acid cycle and hence the energy cycle in some bacteria. The list of antibiotics has grown rapidly over the past few years so that most bacterial diseases can be controlled by their use.

Fungi. Fortunately, there are few pathogenic fungi which attack man, although there are a large number of fungi which are pathogenic to plants. The chemical nature of the fungi protoplasm makes them very difficult to control by chemical treatment. Normally, the host is more affected by the chemical treatment than the fungi, with the net result that chemical treatment has little use except for surface infections. Removal of the diseased portion of the host is the only effective control of fungi disease at the present time.

One of the major reasons why fungi do not attack higher animals more is that they are strict aerobes and must have free dissolved oxygen for growth. Thus it is that major growth can occur only on the body surfaces, in the blood stream, or in the lungs where there is an ample supply of dissolved oxygen. The destruction of plants by fungi is quite common and is of considerable economic importance. Control of the fungi is brought about by the cultivation of fungi-resistant plants and by chemical treatment. Compounds containing sulfur and heavy metals are quite effective in controlling the spread of fungi, provided they are used prior to the infestation.

Protozoa. Very few protozoa are pathogenic to man. The most common pathogenic protozoa of sanitary significance are *Endamoeba histolytica*, *Trypanosoma*, and *Plasmodium*. The protozoa diseases are subject to chemical treatment, but can best be controlled by proper sanitation. Most of the pathogenic protozoa are transmitted to man through a secondary host. The mosquito transmits the *Plasmodium* to man where it continues to grow and upsets the normal metabolic balance. Another mosquito bite of the infected person permits the mosquito to pick up the *Plasmodium* and transmit it to another host.

Toxin Production

Most microorganisms produce a toxic reaction in the host organism by introducing foreign proteins in the form of cellular protoplasm. In a complex organism such as man the foreign proteins result in precipitation of key materials or tying up reactive sites and preventing the necessary metabolic reactions from proceeding. Normally, the toxic reaction due to cellular components is not noticeable until the microorganisms die and release the components forming cellular protoplasm.

A few bacteria, primarily the anaerobic bacteria such as *Clostridium*, produce toxic end products as a result of normal metabolism. *Clostridium tetani* and *Clostridium botulinum* both produce protein end products which cause violent reactions in man. There is very little that can be done once the toxins have been produced, but it is possible to prevent the production of the toxins by proper sanitary precautions.

Pathogen Survival

The pathogens require a host organism for growth and reproduction but can survive for a considerable period outside the host. Water in some form is essential to the transmission of the active vegetative forms of the pathogens, but many of the pathogens have resting stages which permit survival in the absence of moisture. The need for moisture by the vegetative forms prevents their spread far from a host by transmission through the air. Most vegetative forms of the pathogens are spread by poor sanitation through direct contact, water, or food.

On the other hand, pathogens which form spores can travel great distances through the air since dehydration is not a major factor. Most fungi diseases are spread through the air with infection highest in dry dusty areas. Some of the protozoa form cysts which are similar to spores but do not travel well through the air because of their weight. The viruses do not have the moisture content of the vegetative cells of the higher microorganisms and can survive for long periods in the air. It appears that one of the means for transmission of respiratory viruses is through the air. Since spores and viruses do not metabolize significantly, they can survive for long periods outside of the host.

The vegetative pathogens are actively metabolizing whether they are inside or outside of the host organism. Their survival is related in a large part to the quantity of available nutrients and to the environmental conditions. If the environment is unfavorable, the pathogen will die off quite rapidly. With proper environmental conditions the pathogens can survive for long periods. In the heterogeneous biological population in nature the pathogen is faced with considerable competition for nutrients. For the most part since the pathogen must obtain certain key materials from the host organism, it will not compete successfully for the nutrients, but it will obtain sufficient nutrients to survive for an extended period of time.

A good example of survival of a vegetative pathogen is *Salmonella typhosa* in an anaerobic digestion system. *S. typhosa* is an enteric pathogen which has been found extensively in sewage. In the anaerobic digester *S. typhosa* finds an environment suitable for growth as far as temperature and organics are concerned; but it lacks the ability to synthesize tryptophan and depends upon an external source of tryptophan. Normal sewage sludge contains some tryptophan so that *S. typhosa* has a suitable environment for growth in the digester. The normal saprophytic bacteria are able to compete successfully with *S. typhosa* for the tryptophan in the digester so that *S. typhosa* cannot obtain the necessary tryptophan for growth and is unable to grow. Yet *S. typhosa* obtains sufficient tryptophan to sustain life at a very slowly decreasing level for a very long period.

Competition between the parasitic pathogens and the normal saprophytes is probably the major factor in the failure of the pathogens to survive outside of the host organism. It has been proposed that the pathogens do not survive for the lack of a suitable environment or because of toxic materials produced by the other microorganisms. While environmental conditions can retard the growth of the pathogens, most pathogenic bacteria will grow readily under the same environment that permits the saprophytic bacteria to grow as long as sufficient nutrients are present. Most studies have failed to show the presence of toxic end products which are specific for pathogenic microorganisms under normal growth conditions. The antibiotic-producing fungi do not normally produce sufficient concentrations of the antibiotics in soil or water to reduce the bacterial population significantly.

Control of Pathogens

The sanitary microbiologist and the sanitary engineer are interested in preventing the spread of disease by controlling the growth of the pathogenic microorganisms in the mass environment. For the most part control of the pathogens is based on stimulating competition between the pathogens and the saprophytes and adjusting the environment so that the saprophytes are in a better competitive position. The difficulty of chemical treatment on a mass scale makes the use of chemical agents to reduce the pathogens a terminal control method rather than a primary control method.

PART II

Applied Microbiology

CHAPTER 15

Water

In order to survive, all animals and plants must have an ample supply of water free from toxic materials and pathogenic microorganisms. As people congregate more and more in metropolitan areas, the problem of supplying an adequate quantity of pure water becomes greater and greater. One of the prime functions of the sanitary engineer is to ensure that there is always a safe supply of potable water. So well has the sanitary engineer done this job that it is possible to obtain pure, safe drinking water in every town in the United States. The American citizen has become so accustomed to obtaining pure drinking water that he easily falls prey to contaminated water in many areas outside the United States. Like his freedom, the American citizen takes his pure drinking water as a right guaranteed to him under the Constitution.

Source of Water

All water comes in the form of precipitation. It is evaporated from the ocean, condenses to form clouds, and precipitates over land. As the water falls in the form of rain or snow or sleet or hail, it acts as a vacuum cleaner picking up all the dust and dirt in the air. Needless to say, the first water that falls picks up the greatest concentration of contaminants. After a short period of fall, the precipitation is relatively free of microorganisms. When the water hits the ground, a portion of it runs off across the surface of the ground and a portion of it sinks into the ground. The hydrologic cycle is illustrated in Fig. 15-1.

Surface Water

The water running across the surface of the ground has been designated *surface water*. It picks up many substances as it flows back to the ocean, microorganisms, organic matter, and minerals. Surface water collects in low areas forming lakes and ponds, and being rich in nutrients it becomes a perfect medium for the growth of all types of microorganisms. All forms of microbial life are found in surface waters. The types and numbers of microorganisms are a direct reflection of the condition in the water.

If the water is free of minerals, little, if any, biological life will be found. As more organic matter and minerals find their way into the surface water, bacteria, algae, and protozoa grow. Fairly pure waters support few total numbers of microorganisms but have a relatively large number of different species. As more contaminants enter the water, the total number of microorganisms increases, while the number of species

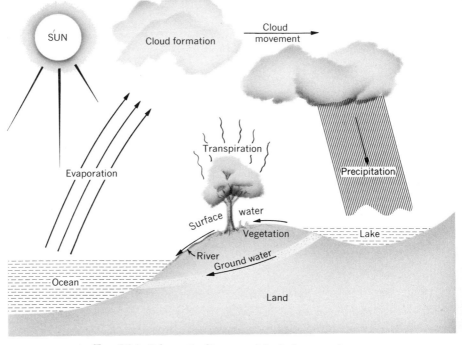

FIG. 15-1. Schematic diagram of hydrologic cycle.

decreases. Waters high in inorganics show excellent algae growth, while waters polluted with organics show predominantly bacteria growths. Surface waters containing large numbers of microorganisms must be treated prior to use. Since an "ounce of prevention is worth a pound of cure," efforts are made to keep most surface reservoirs from becoming contaminated, thereby eliminating the need for extensive treatment.

Ground Water

The surface water which seeps into the ground is designated *ground water*. As it travels through the surface layers of the earth, it picks up some minerals and a few organics in solution. The microorganisms and particulate matter find themselves being filtered out in the upper layers.

Thus it is that most ground waters taken far below the earth's surface are free of microorganisms. These waters are usually relatively low in mineral and organic contaminants. Needless to say, ground waters are usually preferred as sources of drinking water to surface waters.

Pathogenic Microorganisms

Water has long served as a mode of transmission of disease. The disease-producing microorganisms cannot grow in relatively pure water, but they can survive for several days. The vegetative cells die off quite rapidly, but the spores and cysts can persist for an extended period of time. The common pathogenic bacteria found in water include *Salmonella typhosa*, *Shigella dysenteriae*, and *Vibrio comma*. These bacteria have become quite rare in the United States; so much so, that the doctors are not always able to recognize the symptoms of these disease producers.

On the other hand, the viruses have become the prime pathogens carried by water. The small size of the virus and its growth characteristics allow it to escape detection quite easily. In spite of the fact that viruses are not easily detected, no major epidemic in the United States can be attributed to the spread of viruses in the public water systems.

The cysts of certain pathogenic protozoa such as *Endamoeba histolytica* are often found in water supplies in foreign countries. In Mexico the spread of *E. histolytica* is so extensive that it becomes cocktail conversation as to the number of times a person has become infected and taken the cure. Fortunately, this parasite has never obtained a foothold in this country, but many an unsuspecting American tourist has picked up this little parasite in his travels through foreign lands.

Indicator Organisms. In order to control the spread of pathogenic bacteria in water, the sanitary microbiologist must know if the organisms are present. Unfortunately, the analytical procedures for pathogenic bacteria are far from being reliable. Since the pathogens are more fastidious than the common bacteria, they are not detected by normal bacteriological techniques. Rather than look for the specific pathogens, the sanitary microbiologist has found a group of nonpathogenic bacteria whose origin is in fecal matter, and hence is an indicator for the presence of fecal contamination, the coliform bacteria.

The coliform bacteria are members of the family Enterobacteriaceae. They include the genera *Escherichia* and *Aerobacter*. The coliforms were originally believed to be entirely of fecal origin, but it has been shown that *Aerobacter* and certain *Escherichia* can grow in soil. Thus, the presence of coliforms does not always mean fecal pollution. Needless to say, efforts have been made to distinguish between fecal coliforms and the nonfecal coliforms. The differentiation between these two groups is not clear-cut and hence has had limited value. As far as can be determined, *Escherichia coli* is entirely of fecal origin. The intermediate forms of *Escherichia* and the *Aerobacter* are predominantly, but not entirely, of soil origin. Some efforts have been made to determine the presence of *E. coli* as opposed to the other coliforms, but the control of water purity is still based on the presence or absence of any coliforms, soil or fecal forms.

Because of the variation in coliform origins, microbiologists have tried to find other bacteria of fecal origin which were much more specific. The closest bacterial group to meet these specifications are the enterococci. Thus far, the use of enterococci as the indicator organism has not gained acceptance and is still in the experimental stage.

Presumptive Test. The coliforms gained wide acceptance as indicator organisms primarily as a result of the ease with which these bacteria can be detected. The coliforms were found to be one of the few bacteria which could ferment the disaccharide, lactose, with the production of gas. It was possible to detect the coliforms in a sample of water by adding it to a tube of lactose broth containing an inverted vial and incubating at 35°C for 24 to 48 hr. The presence of the peptone in the media stimulated growth of all bacteria; but the coliforms had in the lactose an additional source of food the other bacteria could not use. As the coliforms metabolized the lactose under anaerobic conditions, they formed acids which depressed the pH and forced carbon dioxide out of solution, and assisted in the production of hydrogen gas. These two gases were partially trapped in the inverted vial as evidence of lactose fermentation (Fig. 15-2).

If the coliforms were the only bacteria capable of forming gas in lactose broth, the test would be perfect, but, unfortunately, other bacteria form gas in lactose broth. These noncoliform gas producers include aerobic sporeformers, as well as synergistic action between two or more bacterial species. Fortunately, the noncoliform gas producers occur infrequently. Their occurrence does require that the lactose broth fermentation test be used as a *presumptive* test for coliforms rather than as an absolute test. The microorganisms giving a positive presumptive test must be confirmed as coliforms.

Confirmed Test. Two media are available for the confirmed test, brilliant green bile broth (BGB) and eosin methylene blue agar (EMB). Brilliant green bile broth employs an inhibitor for bacteria growth, brilliant green dye, and a surface tension depressor, bile. The basic medium is lactose broth with a fermentation vial to detect gas production. A small sample from the lactose broth presumptive tube is transferred to the BGB tube. The heavy population in the inoculum allows the coliforms to grow in spite of the inhibitor dye. The other bacteria are unable to grow as rapidly as normal owing to the dye and fail to produce gas within 24 hr of incubation. Aerobic sporeformers cannot grow because the depressed surface tension prevents their obtaining enough oxygen for growth.

With eosin methylene blue agar, two dyes are employed, eosin and methylene blue, in a lactose agar. The inoculum from the lactose broth presumptive test is streaked on the surface of the EMB agar plates in the

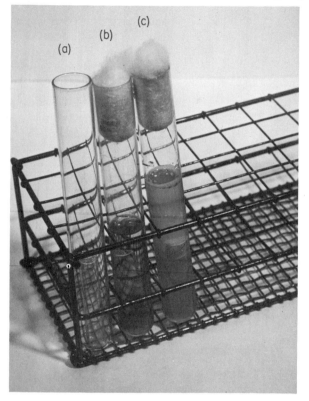

Fig. 15-2. Lactose gas fermentation tubes for coliform analysis: (a) empty tube; (b) uninoculated tube; (c) positive gas formation.

same manner as for isolating pure cultures of bacteria. Most bacteria are inhibited by the dyes, but the coliforms are able to grow because of the lactose. *E. coli* and *A. aerogenes* produce distinctive growths on EMB agar which permits their identification. *E. coli* forms a small, raised, flat, dry colony with a green metallic sheen. *A. aerogenes* forms a large, moist, convex colony with a dark center. The growth of coliforms on EMB agar is shown in Fig. 15-3. The intermediate coliforms have some characteristics of *E. coli* and some of *A. aerogenes*. The ability of EMB agar to distinguish *E. coli* from the other coliforms makes this medium quite popular.

Completed Test. An inoculum from positive BGB tubes or of typical colonies from EMB plates is usually put back into lactose broth to show the microorganisms' ability to ferment plain lactose broth. If the bacteria produce gas, the test is considered positive evidence for the presence of coliforms, and the test is completed.

Most Probable Number (MPN). While it is important to determine the presence of coliform bacteria, it is essential to determine how many coliforms are in a water sample in order to evaluate the extent of contamination. The five-tube MPN test was developed to give an estimate of the number of coliforms in a water sample. The test has performed a good function and has given data which have been successfully used in evaluating fecal pollution. Yet, the test is far from being accurate. It is at

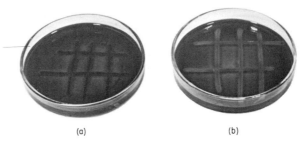

(a)   (b)

FIG. 15-3. Growth of coliforms on the surface of EMB agar: (*a*) *E. Coli*, dark, metallic, dry growth; (*b*) *A. aerogenes*, light, moist growth.

best a statistical evaluation of the most probable number of coliforms, not the absolute number of coliforms. It is amazing how many sanitary engineers take MPN data as absolute data without realizing its crude reliability.

Membrane Filter. The search for a more rapid coliform test and a more accurate test led to the development of the membrane-filter test. The membrane-filter test allows results to be obtained in 24 hr as opposed to a minimum of 72 hr for the completed MPN test. With further developments in media the membrane-filter technique will be simplified and used to a greater extent. The era of prepackaged and presterilized media with disposable dishes is just beginning. There is no doubt that within the next ten years the membrane filter will supersede the MPN technique. The more accurate analyses will permit a better evaluation of water standards and will give absolute results rather than probable results.

Tastes and Odors

While the production of safe water is the prime function of the sanitary engineer, he must also supply a palatable water, free of any strange tastes and odors. Some tastes and odors are strictly of chemical origin, but many have a microbiological origin. It is essential to know which microorganisms cause tastes and odors and how they can be controlled.

Until recently it had been assumed that certain algae were responsible for most microbial tastes and odors, but new studies have implicated the actinomycetes as potential troublemakers. The microbial tastes and odors are caused by certain oils produced by these microorganisms and released on their death. Thus it is with algae, that the tastes and odors do not result until after the algae bloom has occurred and they start dying.

The actinomycetes have been studied so slightly that little is known about their mechanism of taste and odor production. It appears from early research that the actinomycetes live in a parasitic relationship with the algae. The algae are the food supply for the actinomycetes which produce certain taste and odor compounds as normal end products of metabolism. The actinomycetes feed· on the materials released by the algae. When the algae die and lyse, considerable material is released to the solution and rapid growth of the actinomycetes results. The death of the algae and growth of the actinomycetes often mask the true cause of the tastes and odors. In all likelihood both contribute their share to the over-all problem.

The actinomycetes can grow in the absence of algae if sufficient nutrients are available. This can occur in waters containing large quantities of minerals, as well as a source of nitrogen and phosphorus. It is not surprising to see that most of the problems involving actinomycetes have been found in the Southwest and West in areas with highly mineralized water and large agricultural drainage combined with water reuse.

Control of Algae. The control of any biological growth is a function of the nutrients required for energy and protoplasm. Since algae use the sunlight for energy, their control is related to protoplasmic synthesis. The critical element in algae growth is phosphorus. Nitrogen is critical with many algae, but the blue-green algae have the ability to fix atmospheric nitrogen in the absence of soluble nitrogen.

In most natural waters the phosphorus content is very low, and algae growth is limited. But in farming areas where the runoff of surface waters carries away the fertilizers, the phosphorus content becomes quite high and algae growth is proportionally larger. Surface waters receiving the discharge of sewage are likewise fertilized for a good crop of algae. The use of polyphosphate as builders in detergents is likely to increase the algae problems in lakes and reservoirs receiving either treated or untreated sewage unless the treatment process includes phosphate removal.

If it is not possible to prevent the surface waters from being contaminated with phosphate, the next step is to prevent excess growth. Up to a certain population level, the microorganisms do not cause tastes and odors. Thus, the sanitary engineer tries to keep the algae population below the critical population level for tastes and odors. This can be done by the addition of algacides at the right moment. The most common algacide is copper sulfate.

By regular microscopic examination of the reservoir water it is possible to determine the rate of change of the critical microorganisms and to predict when the population is starting to rise. At a predetermined rate of population change the reservoir can be treated before full growth has occurred. The old proverb of "an ounce of prevention is worth a pound of cure" is never more true than in algae control. Addition of copper sulfate to a reservoir before maximum growth can prevent the excess growth with a minimum dosage. Once the algae growth is out of hand, the concentration of copper sulfate required is higher, and when death occurs, the taste and odor materials are still released.

The copper sulfate dosages vary with the chemical characteristics of the water and with the type of algae being killed. In hard waters the copper is precipitated as the carbonate and greater dosages are required. Waters containing organic matter also tie up copper. It is only the free ionic copper that is available for killing algae. The free copper dosage required for algae control ranges from 0.1 to 1.0 mg/liter. Since the algae can only grow near the surface of the water where the sunshine exists, only the upper 5 to 10 ft of the reservoir needs to be treated. The reservoir is treated by distributing the copper sulfate from a boat at a predetermined rate over a set pattern designed to give good distribution.

One of the side effects of copper sulfate treatment is a sudden increase in the bacterial population in the reservoir. Not only is the copper toxic to the algae but also to the predatory protozoa. Death of the predatory protozoa allows the bacterial population, which is unaffected by this copper concentration, to increase to its maximum level. This rise in bacterial population is only temporary, as the protozoa soon become reestablished.

Taste and Odor Treatment. There are times when it is not possible to control the tastes and odors at their source. Various methods are available for treating tastes and odors. Activated carbon, chlorination, and aeration are three of the most common methods used. Activated carbon has tremendous absorption powers for small concentrations of organic matter and is used where chemical coagulation and filtration are employed. Chlorination is popular where little organic matter is in the water. By the use of breakpoint chlorination it is possible to oxidize chemically the taste- and odor-producing substances. Aeration is the simplest method for odor control, but works only if the taste and odor substances are highly volatile at the normal water temperatures.

There is no single solution for taste and odor problems. Each problem is a separate issue which must be evaluated on its own merits. Invariably the final solution is a function of economics and personnel.

Water Treatment

Most ground waters require little, if any, treatment, as they are relatively free of bacteria and other contaminants. This is far from true for surface waters. But in the New England area and in many small areas it is still possible to maintain surface waters in a relatively pure state so that only chlorination is required for treatment. Most surface waters have contaminants such as suspended solids, color, minerals, bacteria, and other microorganisms which must be removed before the water is distributed to the consumer.

The type of water treatment is largely determined by the characteristics of the raw water. Waters high in settleable solids can be stored in large reservoirs until the solids settle out. Waters high in color or colloidal solids need chemical treatment, coagulation, flocculation, and sedimentation followed by filtration. Hard waters are often softened, while high saline waters are demineralized. Usually waters high in bacterial contamination contain high concentrations of solids requiring complete treatment.

Water-quality Standards. Every city and state has some water-quality standards, but the most important set of standards is the USPHS drinking water standards for interstate carriers. The USPHS drinking water standards, while applying only to water carried on airplanes, trains, boats, and buses operating between states, have had widespread implications and have been largely accepted as "the" standards for drinking water, even though their use is strictly for interstate carriers.

The problem of standards is a sore point in sanitary engineering, as many engineers are opposed to setting fixed standards. They feel that once a standard is set, it is impossible to change it. While this has been true in the past, there is no real reason for holding a standard after it has been shown to be no longer useful. Unfortunately, standards have too often become crutches rather than guides. It has become easier to enforce poor standards dogmatically rather than to try to improve the standard to fit current practice.

Raw-water Standards. A good example of rigid standard enforcement lies in the bacteriological requirements for water treatment. To be acceptable for conventional rapid sand filtration followed by chlorination the USPHS recommended that the monthly average density of coliforms not exceed 5,000/100 ml, with not more than 20 per cent of all samples examined exceeding that level. With prechlorination or presedimentation the monthly average density of coliforms must still not exceed 5,000/100 ml but more than 20 per cent of the samples can exceed that level, provided not more than 5 per cent exceed 20,000/100 ml. A recent study by the USPHS has indicated that these standards are low for a modern, well-operated, properly designed water treatment plant. Yet, these standards remain and will remain until the USPHS or control authorities revise them in the light of new data. Our rapidly changing technology makes it imperative that we examine our standards at frequent intervals to make sure that they are workable, realistic standards which can be used as flexible guides for evaluating each particular situation rather than as fixed rules regardless of the situation.

Finished-water Standards. The USPHS drinking-water standards for interstate carriers set the minimum number of samples to be collected, as well as the bacterial quality to be obtained. The frequency of sampling for bacteriological analyses is a function of the population served by the particular distribution system at hand.

Population served	Min. no. samples per month
0–2,500	1
10,000	7
25,000	25
100,000	100
1,000,000	300
2,000,000	390
5,000,000	500

These are only the routine samples and do not include special samples run to check contamination.

Because of the very low density of coliforms in finished waters it is not necessary to run the normal three series of five tubes for the MPN. Only one set of tubes is required. This is where the membrane filter has an advantage since any volume of water can be used. Two standard samples have been used, a 10-ml and a 100-ml. To meet USPHS requirements not more than 10 per cent of all 10-ml samples examined per month can show the presence of coliforms. Occasionally, three tubes of the five-tube 10-ml test can show coliforms, provided this does not occur in consecutive samples or in more than 5 per cent of the samples when 20 or more samples are examined monthly or in more than 1 sample when less than 20 samples are examined. When three or more positives are obtained on a single sample, daily samples must be taken until at least two consecutive samples show the water to be satisfactory.

With 100-ml samples, not more than 60 per cent can show the presence of organisms of the coliform group; but occasionally all five portions may be positive provided it does not occur in consecutive samples or in more than 20 per cent of the samples when five or more samples are examined per month or in one sample when less than five samples are examined. When all five portions of a single sample show the presence of coliforms, daily samples are examined until at least two consecutive samples show satisfactory results.

Storage. One of the simplest methods for water treatment is storage in a large reservoir. Under quiescent conditions the heavy suspended particles settle out. Bacteria are absorbed by many of these settleable particles and are removed by sedimentation also. Studies in the TVA reservoirs showed 80 per cent coliform reduction for 50 per cent of the time before and after impoundment for Knoxville raw water. Other studies have shown as much as 99 per cent coliform reduction between the influent water and effluent water from Douglas Reservoir.

It is well known that a bacteriological sample stored for a long period will decrease quite rapidly. A 5-day storage period can result in a coliform decrease of as much as 90 per cent. The application of this phenomenon in practice can result in a definite coliform reduction at a low cost where land is available.

Recently, Dallas, Tex., was faced with a water shortage. Two sources of water were available, Trinity River water and Red River water. The Trinity River water flowed right by Dallas but contained the treated sewage from Fort Worth and adjacent communities. The Red River water was over 30 miles away and was a very highly mineralized water. One of the leading consulting engineering firms in Dallas was retained to make a proper evaluation of both water sources. The engineers recommended 60-day storage of Trinity River water followed by complete treatment. Such a storage would have made the water bacteriologically better than the water they were currently using. Unfortunately, politicians overrode the engineers' decision and chose the Red River water at a considerable economic loss to the citizens of Dallas and a definite deterioration in water quality. The only consolation to the engineers was that they at least made the right evaluation even if it was ignored.

Chemical Treatment. One of the most widely used forms of water treatment is chemical treatment with alum. In most natural water containing colloidal particles of bacteria, clay, or organics, alum forms an insoluble precipitate with a positive electric charge due to an excess of aluminum ions which are strongly positively charged. The colloidal particles mostly carry a negative electric charge at normal pH levels and are attracted to the positively charged alum floc. The combination of positive- and negative-charged particles lowers the net charge and allows the particle to grow quite large when slowly agitated in a flocculator. Under quiescent conditions the large floc particles settle out and remove a considerable portion of the bacteria. The efficiency of bacteria removal varies quite widely, being largely a function of the sedimentation tank. With a well-designed flocculation and sedimentation system it is possible to obtain 90 to 99 per cent bacteria removal. Some of the new solids contact systems are very effective in reducing bacterial populations. The efficiency of the newer systems has reached the point where some engineers are questioning the need for the sand filter following sedimentation.

Sand Filtration. One of the major advances in sanitary engineering was the sand filter. Originally, sand filtration was complete in itself. The raw water was applied to the sand filter and filtered at a very slow rate. The particulate matter was strained out near the surface of the filter and gave a source of nutriment to microorganisms. As the microorganisms grew, they formed a very fine film which greatly assisted in removing both colloidal and soluble organic matter from the raw water. This surface film was known as the "smutzdecke" and was essential to slow sand-filter efficiency. Over 99.99 per cent of the bacteria in the raw water were removed by a well-operating slow sand filter.

The length of operation of the slow sand filter was a function of the material removed from the raw water. It became obvious that the filter runs could be increased if a large part of the suspended solids were removed prior to filtration. It was at this point that sedimentation came into use before filtration. Chemical treatment improved the efficiency of sedimentation and greatly extended the filter runs. It was possible to increase the rate of filtration without impairing effluent quality too materially, resulting in the modern rapid sand filter with its 2-gpm/sq ft rate. Better sedimentation efficiency has prompted research into increased filter rates. Chicago has found that a 4-gpm rate does not result in much impairment in effluent quality. With more efficient methods of chlorination it is possible to effect the same degree of treatment as before, but in a shorter time period.

With increased treatment efficiencies the need for sand filters has become less and less. It appears that the sand filter has almost outlived its economic value in many areas of the United States. The cost of filtration in terms of its effectiveness is rapidly exceeding other forms of treatment.

Chlorination. The final step in conventional water treatment is chlorination. The filtered water contains few organisms and little organic matter. As a result it is possible to add a small concentration of chlorine to remove effectively all pathogenic bacteria. An applied dosage of 0.1 mg/liter of chlorine will usually give complete kill of all bacteria of sanitary origin. The effectiveness of chlorination is a function of time, pH, temperature, bacterial population, and residual chlorine.

In order to give the chlorine time to destroy the bacteria, the chlorine is usually added immediately after filtration before the clear-water storage. Retention in the clear-water storage allows the chlorine to react with the microorganisms.

The effectiveness of a given chlorine dosage is greater at pH 7.0 than at pH 9.0. This is believed due to the fact that in water chlorine reacts with water to form hydrochloric acid and hypochlorous acid.

$$Cl_2 + HOH \rightarrow HCl + HOCl \qquad (15-1)$$

The hypochlorous acid is the key material in bacterial toxicity. As the pH increases, the hypochlorous acid undergoes dissociation.

$$HOCl \rightarrow H^+ + OCl^- \qquad (15-2)$$

It appears that the undissociated form is much more active than the dissociated form. Thus, as the pH increases, the dissociated form increases with a resultant decrease in bactericidal efficiency. Chlorine is effective in destroying microorganisms because it oxidizes key reducing enzyme systems and prevents normal respiration. At high chlorine dosages denaturation of the protein results with complete destruction of the cell. The latter usually does not occur until after breakpoint chlorination residuals have been obtained.

The rate of all chemical reactions are affected by temperature, and chlorination is no exception. At $0°C$ chlorine reacts extremely slowly. As the temperature increases, the rate of reaction increases very rapidly, giving almost twice the rate of kill for each $10°C$ rise.

One of the most interesting concepts in sanitary microbiology has been the misuse of "per cent kill." It is quite easy to speak of 99 per cent kill or 99.99 per cent kill. It makes the treatment process sound very effective even if it is not. A 99 per cent kill with an initial bacterial population of 100 means one cell remains, but the same per cent kill with an initial population of 10,000 means 100 bacteria remain. Both kills sound good, but the residual bacteria count in the second case is far from satisfactory. In considering the effectiveness of chlorination the only sound criteria are "residual bacteria counts." The safety of the water system from bacterial contamination is not a function of the per cent kill but of the residual bacterial population.

Needless to say, the chlorine dosage will affect the rate of kill. If storage is available for a long contract period, the chlorine dosage for effective bacteria kill can be much less than if limited storage is available.

Softening. In hard-water areas an additional softening treatment is often included in the over-all water treatment process. Softening consists in adding sufficient quantities of lime and soda ash to precipitate the calcium and magnesium as calcium carbonate and magnesium hydroxide. In order to precipitate the magnesium, it is necessary to raise pH above 10 to yield the free hydroxide ions. Free hydroxide ions have already been shown to be toxic to bacteria so that it is not surprising to find that softening plants usually produce better bacteria kills with less effort than nonsoftening water treatment systems.

Distribution System

Once the water has been treated satisfactorily, it must be delivered to the consumer. This is done in the distribution system. In order to prevent bacteria contamination which might accidentally enter the distribution system, it is standard practice to keep a definite chlorine residual throughout the distribution system. This is not easy to do, as the chlorine reacts with the iron pipe and is reduced to chloride which has no effect on the bacteria. Often a heavy chlorine dose is required at the treatment plant in order to produce a chlorine residual at the far end of the distribution system. This usually leads to complaints of excess chlorine from consumers near the treatment plant. As a result, the chlorine dose is reduced below the complaint level with the net effect that the far end of the distribution system goes unprotected and biological growths result.

For the most part, biological growths occur in dead ends where the low concentration of organic matter allows a slow accumulation over a long period of time. Fortunately, the organic concentration and nutrient concentration is sufficiently low that biological growth is quite slow. One of the most common forms of biological growth is the autotrophic iron oxidizing bacteria. These bacteria oxidize the ferrous iron to ferric iron while utilizing carbon dioxide as their carbon source. These bacteria are usually found where corrosion is rapid and may even assist in the corrosion of iron pipes. The recent use of polyphosphates for corrosion prevention in water distribution systems supplies a key nutrient long absent. It will be interesting to see if increased biological growths result in distribution systems treated with polyphosphates. Fortunately, nitrogen will still be low in most treated waters and will limit growth. The easiest way to eliminate biological growths in distribution systems is always to use loops so that the growths are continuously flushed out before they even become large enough to see and to maintain a chlorine residual in all parts of the distribution system.

Virus. Control of bacteria-caused disease has increased the number of virus-caused diseases. The increased concentration of viruses in sewage effluents means the potential concentration of viruses in sewage-contaminated surface water is going to increase. Thus far, viruses have not been implicated as being spread through water distribution systems. But it is important to realize their implication before epidemics result.

Studies have shown that normal water treatment processes are relatively effective in reducing viruses but not as effective as for bacteria.

Chlorination is not as toxic to viruses because of the simpler structure and the need for denaturation of the nucleoprotein, which requires a high concentration of free chlorine. The free chlorine residual to kill viruses is about twice as high as for bacteria.

One of the major problems with viruses is their detection. Virus growth is at present a complex procedure which cannot be run routinely. It has been shown that the normal coliform index can be used as a rough guide for enteric viruses, as well as for enteric bacteria. This is fortunate for the sanitary microbiologist, but it is apparent that more and more attention will have to be paid to viruses, especially as sewage effluents are reused for drinking water.

SUGGESTED REFERENCE

1. "Standard Methods for the Examination of Water and Wastewater," 11th ed., American Public Health Association, New York, 1961.

CHAPTER 16

Liquid Wastes

One of the most important uses of water is for removal of waste materials. In cities and small suburban areas the domestic and industrial wastes are removed with water in extensive collection systems. It is one of the major functions of the sanitary engineer to design satisfactory systems in which the liquid wastes are collected and disposed of without creating a health hazard or nuisance conditions. There are two major groups of liquid wastes, domestic sewage and industrial sewage. The characteristics of these two sewages are quite different and will be discussed separately.

Domestic Sewage

The household wastes from a purely residential area are defined as domestic sewage. Strangely enough, the characteristics of domestic sewage are relatively constant. This fact has greatly assisted the sanitary engineer in the design of sewage collection and sewage treatment systems and at the same time has created problems with the exceptions.

The domestic sewage from one residential area is approximately the same as from any other residential area because of the habits of the people within a region or a country. In the United States the pattern of daily activities begins with rising between 6 and 7 A.M., to work between 8 and 9 A.M., coffee break at 10 A.M., lunch between 12 and 1 P.M., coffee break at 3 P.M., home between 4 and 5 P.M., dinner at 6 or 7 P.M., and bed by 11. This routine sets the pattern of sewage flow and strength (Figs. 16-1 and 16-2). The type of food, much of which is prepackaged, frozen, and in cans, is relatively uniform and sets the characteristics of the domestic sewage.

The changing technology of this modern civilization has produced some rapid changes in sewage characteristics and will probably produce more in the years to come. It is essential that the sanitary engineer be fully aware of the past changes in order that he might be able to predict future changes. The most radical changes in the past ten years have been

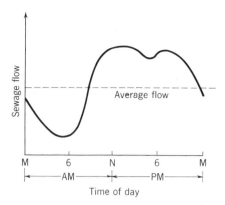

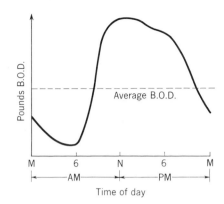

FIG. 16-1. Variations in sewage flow during a typical day.

FIG. 16-2. Variations in B.O.D. during a typical day.

the increase in garbage grinders and in the replacement of soap by synthetic detergents. Garbage grinders have had two effects on sewage characteristics, increasing the total organic load and increasing the hydraulic flow. Synthetic detergents, commonly referred to as syndets, have caused the replacement of a single organic compound by a multitude of unusual chemical structures. While soaps undergo one series of chemical reactions in sewage, the syndets follow a different series.

Present Sewage Characteristics

Examination of the average sewage flows from residential districts with moderately well-constructed sewers will show 50 gal per capita per day, provided storm drains are not connected to the sanitary sewer. If the suburban dwellings contain automatic clothes washers, automatic dishwashers, automatic garbage grinders, and two baths with showers, the sewage flow can rise up to 75 gal per capita per day.

The organic contribution of sewage is measured in terms of 5-day biochemical oxygen demand (B.O.D.). The average per capita contribution of B.O.D. is 0.15 lb/day, making the average contribution of B.O.D. based on 50 gal per capita per day 360 mg/liter.

A residential community with normal businesses and light industries will average 100 gal per capita per day with a normal 5-day B.O.D. contribution of 0.2 lb per capita per day. This will give an organic concentration of 240 mg/liter 5-day B.O.D. The nature of the industries will cause the volume of sewage and its characteristics to vary quite widely from the above averages. Each community with large industries requires careful analyses to be made of sewage flow patterns and characteristics, as there are no averages which can be used.

Inorganics. Domestic sewage contains both inorganic and organic compounds. The inorganic characteristics are determined in part from the carriage water and from infiltration. This will lead to wide variations in inorganic concentrations. Well-constructed sewers will have sewage with lower inorganic concentrations than sewage in poorly constructed sewers. It is possible to use certain inorganic ions to determine the extent of infiltration of ground water into the sewers. Any ions which have a significant differential between the sewage and the ground water can be used to estimate infiltration. The inorganic ion concentrations in sewage range from 300 mg/liter to 3,000 mg/liter, with an average of 500 mg/liter. The approximate distribution of inorganic ions by weight is as follows:

Inorganic ions	Per cent
Sodium	25
Potassium	2
Ammonium	4
Calcium	5
Magnesium	1
Iron	Less than 1
Bicarbonate	40
Sulfate	10
Chloride	10
Phosphate	1

Organics. The organic fraction of domestic sewage averages approximately 300 mg/liter, dry weight, with a range from 100 to 500 mg/liter. In fresh sewage approximately 80 per cent of the organic matter is insoluble, either settleable or colloidal. As sewage ages, biological action converts more of the insoluble organics to soluble organics. In stale sewage 50 to 60 per cent of the organics will be soluble.

The organic matter in sewage can be divided into three major groups: proteins, carbohydrates, and fats. Of the 300 mg/liter organic solids, 40 to 50 per cent is protein, 40 to 50 per cent is carbohydrate, and 5 to 10 per cent is fat. The proteins are complexes of amino acids which form the major source of microbial nutrients. The carbohydrates can be divided into two groups: (1) the readily degradable starches and sugars and (2) cellulose. The starches and sugars are easily metabolized by the microorganisms in sewage, while the cellulose compounds are degraded at a much slower rate. The fats are not very soluble and are degraded by microorganisms at a very slow rate. The replacement of soaps by synthetic detergents has resulted in a decrease in the total fat content by less than 10 per cent. Since the syndets are similar in structure to soaps and other fatty compounds, they will be considered as part of the fatty fraction.

Of the organic matter in sewage it appears that between 20 and 40 per cent is nonbiologically degradable. Very little is known about the non-degradable portion of sewage since it does not exert an oxygen demand and hence has little effect on the sanitary properties of sewage. A portion of the nondegradable organics is soluble and will remain in the liquid phase.

Microorganisms. Sewage is a perfect microbiological medium with all the inorganics and organics necessary for good microbial growth. Some of the microorganisms enter the sewage at its source, but many come from the soil by infiltration. It is possible to find almost every type of micro-organism in sewage.

Needless to say, the numbers and types of microorganisms in sewage will be largely a function of the environment. In fresh sewage there will be approximately 10^5 bacteria per milliliter while in stale sewage the bacteria count will rise to 10^7 or even 10^8 bacteria per milliliter. The strict aerobic bacteria cannot grow in stale sewage because of the absence of dissolved oxygen; but they are present in the dormant spore stage and are ready to grow as soon as the environment becomes satisfactory.

The other aerobic microorganisms such as fungi and protozoa are also in sewage, primarily in the dormant spore or cyst stages. At the sewage-air interface there is often enough oxygen to allow some of the lower forms of fungi and protozoa to grow. These microorganisms normally attach tnemselves to the sewer and thus are able to remain in an aerobic environment while the rest of the sewage is strongly anaerobic.

Collection of Sewage

Sewage is collected by a sewerage system and is conveyed to a point of discharge. In a large community the sewerage system can become quite extensive. For the most part the sewage flows by gravity from the individual houses in small sewers into larger sewers which in turn empty into larger interceptor sewers. In some areas it is necessary to lift the sewage at various intervals by means of pump stations. As the sewage flows along, biological action is progressing at an ever-increasing rate. The longer the sewage remains in the sewerage system, the greater will be the biological action. If the sewers are very long and the rate of flow is low, the biological action will become so extensive that the sewage can be considered as receiving partial biological treatment. In the South-west where the sewer slopes are quite flat and the temperature becomes quite high, many cities receive primary biological treatment in the sewers. While this action offers certain advantages, it has definite disadvantages, namely, odors, septicity, and corrosion of the sewer pipes.

Sewage Odors

Sewage odors arise from the anaerobic degradation of proteins and the reduction of sulfates. The anaerobic degradation of porteins is brought about by the facultative bacteria and the strict anaerobic bacteria. The end products of their metabolism include skatole, mercaptans, butyric acid, and miscellaneous aldehydes, all of which are odorous, volatile compounds which escape as gases from the sewage.

Under normal circumstances with sewage flows sufficiently rapid to prevent deposition of solids, the only odors will be from protein decomposition. If the nature of the sewage is such that it is high in fats and the flow is low, the solids in the sewage will tend to adhere to the sewer wall in fatty clumps. The organic deposits allow the microorganisms to grow up at a single point over a prolonged period. Under anaerobic conditions the normal facultative bacteria lack sufficient hydrogen acceptor to degrade the fats, permitting the fats to remain for a long period of time. The sulfate-reducing bacteria are able to attack the fats since they have a hydrogen acceptor which is unavailable to the other bacteria, namely, sulfates. As the sulfate-reducing bacteria metabolize the fats, hydrogen sulfide is produced as an end product. When the metabolism becomes sufficiently rapid, large quantities of hydrogen sulfide are produced, and it goes off as a gas into the atmosphere above the sewage.

There are many ways to control sewage odors and to retard septicity. The most effective odor control measure is to prevent microbiological activity from reaching the stage where the concentration of odorous end products is detectable. This can be done by proper sewer design, regular cleaning, or chemical treatment. In designing a sewerage system, the most important items are retention time of the sewage in the sewer and the scouring velocity.

If sewage is not retained in the sewers for a long period, the metabolic activity of the microorganisms will not be sufficient to create strong anaerobic conditions and odorous end products. Since temperature is an important factor in metabolic activity, it is essential that the retention period be shorter in warm climates than in cool climates. In the Southwest where the temperature is high and the terrain flat, odor problems are quite prevalent. Since neither the retention period nor the temperature can be changed, chemical treatment must be used for odor control.

The primary form of chemical treatment is through the use of a strong oxidizing agent such as chlorine. Unfortunately, chlorine is so reactive that it ties up the organic matter in the sewage and prevents its reaction with the microorganisms. In order to "tame" the chlorine, it has been combined with hydrocarbons to form chlorinated hydrocarbons which retain some of the microbial toxicity but do not react with other organics.

The reactivity of the chlorinated hydrocarbons is low compared with chlorine. As a result, they have not always produced the desired effects. The prime advantage of the chlorinated hydrocarbons is absorption to grease deposits, where they prevent the growth of the *Desulfovibrio*. When the odors are due to protein decomposition by the normal facultative bacteria in the flowing sewage, the chlorinated hydrocarbons are not very effective. Some of the bacteria in sewage can metabolize the chlorinated hydrocarbons the same as any organic compound, thereby removing them from the sphere of reaction. The difference in activity of the chlorinated hydrocarbons may help explain their success in some instances and their failure in others.

Corrosion of Sewers

The production of hydrogen sulfide in sewage is usually accompanied with a depressed pH as a result of acid formation by anaerobic metabolism. Acid conditions allow the hydrogen sulfide to exist in the unionized state and to be easily volatilized. The gaseous hydrogen sulfide is partially soluble in the moisture of condensation on the sewer walls. The moisture of condensation on the side walls usually drops back into the

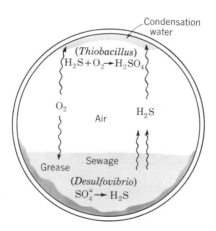

FIG. 16-3. Schematic diagram of H₂S corrosion in sewer pipe.

sewage, but the moisture on the crown remains for long periods. The sulfur-oxidizing bacteria, *Thiobacillus*, find the hydrogen sulfide environment satisfactory and oxidize the hydrogen sulfide to sulfuric acid. The *Thiobacillus* can carry on metabolism at pH levels as low as 1.0. In concrete sewers the sulfuric acid which has been formed reacts with the lime to form calcium sulfate which lacks the structural strength to hold the sewer together. The occurrence of crown corrosion has become so extensive in warm climates that acid-resistant vitrified clay pipes are used for all but the largest sewers. The H_2S corrosion cycle is shown in Fig. 16-3.

Needless to say, the best control for crown corrosion is the prevention of excess hydrogen sulfide. The next best control is adequate ventilation in the sewer to prevent moisture from condensing on the crown and to keep the hydrogen sulfide concentration in the sewer atmosphere at a minimum. The use of lime to form calcium sulfide instead of hydrogen sulfide has been tried, but the formation of heavy, insoluble calcium salts can result in increased deposits of organic matter in the sewer. The quantities of lime required to control hydrogen sulfide loss to the atmosphere are usually so great that this is not a widely used solution.

Biochemical Oxygen Demand Test

One of the most important tests for the determination of the organic strength of sewage is the 5-day B.O.D. test. Essentially, it is the measure-

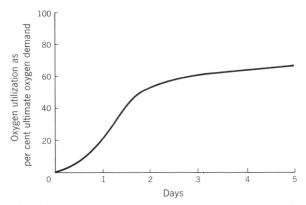

Fig. 16-4. Oxygen utilization during 5-day B.O.D. test.

ment of the oxygen utilized in the stabilization of the organic matter in sewage by microorganisms over a 5-day period at 20°C. With domestic sewage the 5-day B.O.D. values represent 65 to 70 per cent of the total biologically oxidizable organic matter and is a relatively consistent value (Fig. 16-4).

The 5-day B.O.D. test evolved from the fact that in England none of the streams had more than a 5-day flow period. Thus, the 5-day B.O.D. was the maximum oxygen demand which would be exerted in any stream in England. Transportation of the B.O.D. test to the United States lost most of its original significance since many streams had more than 5 days' flow to reach the ocean. In spite of the lack of significance of the 5-day incubation period in the United States, it has been retained even though shorter incubation periods have been shown to yield as accurate results.

The B.O.D. test, although used as a quantitative test, is actually a qualitative test with a narrow range of variation that permits it to assume semiquantitative proportions. Its long period of use has caused many sanitary engineers to forget that the B.O.D. value is not an absolute measure. The prime factor in the B.O.D. test is the presence of micro-organisms in the sewage. The retention of sewage for a short period in the sewers allows the bacteria to begin to degrade the organic matter and to produce additional bacteria. The bacteria which grow are adapted to all the components of sewage and represent a thoroughly acclimated microbial population.

In fresh domestic sewage with 240 mg/liter 5-day B.O.D. and 1×10^5 bacteria per milliliter, a 1:40 dilution would be the lowest used for the B.O.D. test. This would give approximately 3×10^3 bacteria per milliliter and 9 mg/liter ultimate B.O.D. The high food-microorganism ratio permits unrestricted bacteria growth in the log phase. The bacteria will follow the log phase until food becomes limiting. In this case approximately 50 per cent of the organic matter will be stabilized at the end of the low growth phase, which will take approximately 24 hr. Stabilization of this fraction of organic matter will exert approximately 1.8 mg/liter oxygen demand and will produce approximately 1×10^7 bacteria per milliliter. The lack of food slows down bacteria growth, but a second factor becomes even more important, protozoa growth. The free-swimming ciliates find the bacteria a perfect substrate for growth. As soon as the bacteria population reaches reasonable levels, the protozoa begin to grow. It takes approximately 1×10^5 bacteria to produce a protozoa so that the protozoa growth will definitely lag the bacteria growth. There is little change in the protozoa population during the first 24 hr. The second 24 hr sees the bacteria being held back on the growth curve by the protozoa, reducing M and keeping the $F:M$ ratio higher than would be encountered if only the bacteria were present. Essentially, all the remaining food is removed with the production of an additional 6×10^6 bacteria per milliliter and an exertion of 2.7 mg/liter oxygen demand. The protozoa population has grown from 1 or 2 per milliliter to 30 per milliliter. The protozoa have consumed 3×10^6 bacteria per milliliter in the process of growing and have exerted an oxygen demand of 0.3 mg/liter. Thus at the end of 2 days the bacteria population stands at 1.3×10^7/ml and the protozoa population is 30 per milliliter, with the total oxygen demand at 4.8 mg/liter.

The third day finds the bacteria in endogenous metabolism but decreasing as a result of the protozoa growth. The protozoa population probably reaches 100/ml, which decreases the bacteria population to

6×10^6/ml. Endogenous metabolism by the bacteria alone drops the bacteria to 4×10^6/ml. The protozoa exert an oxygen demand of 0.7 mg/liter, while the bacteria exert 0.5 mg/liter. The net effect is a 5.5 mg/liter oxygen demand by the end of the third day.

The protozoa population probably holds at 100/ml in the endogenous phase which requires 1×10^6/ml bacteria. Bacterial endogenous metabolism will drop the population another 1×10^6/ml. This yields a total bacteria population of 2×10^6/ml and an addition of oxygen uptake of 0.25 mg/liter to give a total of 5.8 mg/liter B.O.D.

The important day of the B.O.D. test, the fifth day, finds all microorganisms decreasing in numbers. The protozoa drop to approximately 50/ml, while the bacteria drop only 5×10^5/ml to 1.5×10^6/ml. The

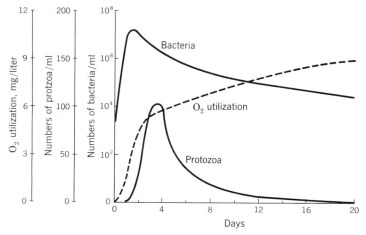

FIG. 16-5. Changes in bacteria, protozoa, and oxygen during 20-day B.O.D. test.

total oxygen uptake rises to 6.0 mg/liter, 67 per cent of the ultimate B.O.D. The bacteria and the protozoa continue their endogenous metabolism with almost complete metabolism by the twentieth day. The changes in the bacteria, protozoa, and oxygen demand during the B.O.D. tests are shown in Fig. 16-5.

During the 5-day period the bacteria have exerted 4.8 mg/liter oxygen uptake, or 80 per cent of the oxygen demand, while the protozoa have accounted for the remaining 20 per cent. The protozoa not only exert their own oxygen demand but stimulate the bacteria to a greater demand per unit of time. The total oxygen potential of any waste is a fixed value with the only variable being time required to exert that demand. If bacteria alone were stabilizing the entire organic matter, the time for complete stabilization would be extended out several more days. The metabolism of bacteria by protozoa actually results in a greater demand for energy per unit of time and hence oxygen.

Microbial Seed. Variations in the initial bacteria populations will have little effect on the 5-day B.O.D. as long as the numbers exceed 1×10^3/ml. Less than 1×10^3 bacteria/ml delays the bacterial growth quite con-

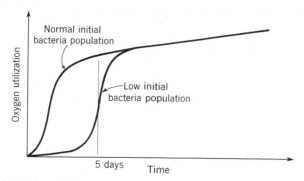

FIG. 16-6. Effect of low initial bacteria population on 5-day B.O.D. test.

siderably and hence the protozoa growth (Fig. 16-6). A very low bacteria population can result in loss of the protozoa activity during the first five days so that the 5-day B.O.D. value would be below 67 per cent.

If the bacteria are not adapted to the organic matter in the B.O.D. bottle, they will decrease in numbers until all the bacteria are dead or

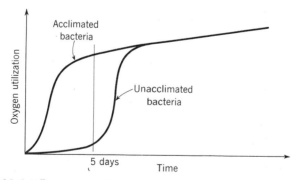

FIG. 16-7. Effect of unacclimated bacteria seed on 5-day B.O.D. test.

until the bacteria adapt to the substrate. The net result is a lag followed by the normal B.O.D. curve (Fig. 16-7). The 5-day B.O.D. value for an unacclimated seed will be quite low compared with an acclimated seed. It is important that acclimated seeds be used to obtain true values of the organic strength. Often control authorities fail to obtain correct

B.O.D. values on industrial wastes, as they do not use acclimated seed. Since the B.O.D. values will be low, the industry never complains.

An acclimated seed can be produced by aerating sewage together with the particular waste to form an activated sludge. The activated sludge will have all the necessary bacteria and protozoa for complete metabolism of the organic matter in the wastes. Production of an acclimated seed on certain exotic organic compounds such as those currently being synthesized in industry and appearing in industrial wastes is not an easy task and may take many months. Acclimation in such cases is produced by growing the microorganisms on organic compounds with similar chemical structures but which are readily degradable. Once the bacteria are metabolizing the easily degradable compounds, the hard to metabolize compound is substituted slowly while the easily degradable compound is withdrawn. In this way the microorganisms are forced over to the difficult compound. Studies with sulfanilic acid took over six weeks to develop a proper microbial seed.

Nitrification. With a domestic sewage seed, nitrification will set in on the B.O.D. test between the fifth and seventh day under normal circumstances. Since the B.O.D. test is determined on the fifth day, nitrification is normally not a problem. Highly stabilized effluents from secondary biological treatment systems contain a high population of nitrifying bacteria. The nitrifying bacteria will attack any unoxidized nitrogen immediately, creating an oxygen demand due to the nitrification reaction. Nitrification can be reduced by pasteurizing the seed before placing it in the B.O.D. bottle or by the addition of sulfuric acid and subsequent neutralization with sodium hydroxide.

Algae. Like nitrification, algae have been responsible for certain problems in the B.O.D. test (Fig. 16-9). In the presence of light the algae

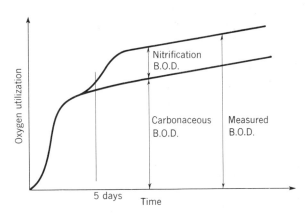

Fig. 16-8. Effect of nitrification on 5-day B.O.D. test.

will often produce more oxygen than the bacteria consumed. The net effect is an increase in dissolved oxygen in the B.O.D. bottle. In the dark the algae exert an oxygen demand just like bacteria and other micro-organisms. If the algae die, they lyse and release their nutrients into

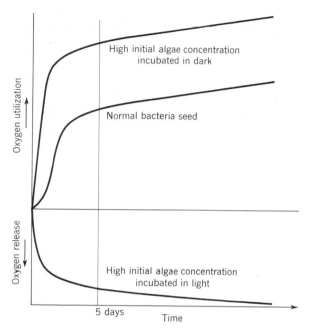

FIG. 16-9. Effect of algae on 5-day B.O.D. test.

solution; the bacteria can utilize these nutrients and will require more oxygen than if the algae were absent. Since the rate of oxygen demand by the algae is low, the B.O.D. due to the algae will be minor. It is reasonable to determine the B.O.D. of algae containing waters in the dark with the realization that the B.O.D. values thus obtained will be higher than those actually occurring in receiving waters.

CHAPTER 17

Industrial Wastes

The wastes from industries consist of sanitary sewage, process wastes, and cooling waters. These wastes are not like domestic sewage and pose special problems of their own. The strength and composition of industrial wastes vary considerably from domestic sewage and even from industry to industry or with the same type of plant within a single industry. This variability in industrial wastes has created many problems for the sanitary engineer who normally has been used to the relative uniformity of domestic sewage.

Sanitary Sewage

The sanitary sewage from an industry is different from domestic sewage. In a light office building 90 per cent of the domestic sewage will come from toilets and wash basins, while the remaining 10 per cent will come from water fountains and weekly floor washings, if necessary. This sewage comes within the normal 8- to 10-hr working period. The average sanitary sewage volume for a light office building will be 20 gal per person per shift, with about 0.05 lb 5-day B.O.D. per person per shift.

With light manufacturing the sewage flows will go up to 25 gal per person per shift, with 0.06 lb 5-day B.O.D. per person per shift. As the manufacturing operation becomes more complex, there is a need for regular shower facilities and considerably more clean-up. This can run the domestic sewage contribution to 40 gal per person per shift, with 0.10 lb of B.O.D. per person per shift. Extreme care must be used in estimating the strength and volume of sanitary sewage from industry. It is amazing how few sanitary engineers can make a reasonable estimate of sanitary sewage from an industry. The attitude is basically one of estimating the sewage quantity and then doubling it to be safe.

Process Wastes

Each industry produces its own characteristic process wastes. These wastes result from washing operations: washing of the raw materials, the intermediate product and the final product, as well as washing the product containers. The wash waters usually contain a small quantity of the process materials, which imparts the characteristics to the process wastes. For the sake of convenience the process wastes will be divided into two sections, inorganics and organics.

Inorganic Process Wastes. Inorganic process wastes result primarily from the metallurgical industries, the inorganic chemical industry, and the petroleum industry. In the cleaning of steel, large quantities of sulfuric acid are used. The acid rinse waters contain large quantities of sulfuric acid and iron sulfate. The spent acid rinse waters have an adverse affect on the microbiology in sewage or in streams and must be treated to remove the mill scale and to neutralize the acids before being discharged to sewers or to a normal watercourse.

The metalplating industry uses large quantities of acids such as sulfuric, hydrochloric, and nitric and toxic chemicals such as chromates and cyanides. In addition to these chemicals appearing in the rinse-water baths some of the metallic ions such as silver, copper, and iron are also present. The rinse waters from the metalplating industries require extensive treatment either by large dilution with domestic sewage or by chemical treatment to reduce the chemical concentration below the toxic level.

The processing of inorganic chemicals from raw ores results in the discharge of wash waters containing unwanted materials. The contaminant concentration is often so great as to make disposal by dilution almost impossible. To prevent damage to the biological life in watercourses, it is necessary to lagoon the wash waters and allow solar energy to evaporate the water, leaving the unwanted materials in the solid state.

The petroleum industry produces tremendous quantities of waste salt solutions which must be disposed of properly to prevent land damage, as well as water damage. These waters have been pumped back into the ground and have increased the oil yields, as well as solving a pollution problem.

For the most part, the inorganic wastes do not pose many biological problems other than toxicity. When combined with organic wastes, they can accentuate the problems of the disposal of the organic wastes. A high sulfate concentration with fatty acids will stimulate the sulfate-reducing bacteria and create a hydrogen sulfide problem.

Organic Process Wastes. The major pollution problem from industrial wastes lies in the disposal of the organic process wastes. The major sources of organic wastes are the food-processing industries, the paper industry, the textile industry, the petroleum industry, and the chemical industries.

Food-processing Wastes. The canning of food, the processing of sugar, the production of milk products, and the preparation of meat products are all major waste-producing operations. Food canning requires the use of large volumes of water for washing prior to canning and cleaning up after canning. Both of these operations result in the production of large quantities of organic wastes which are easily degraded biologically. Sugar and syrup processing also results in spent waste waters with a high carbohydrate content. Cleaning operations alone in both the milk and the meat operations are responsible for large quantities of high organic waste waters.

Roughly speaking, the 5-day B.O.D. of food-canning wastes can range from 500 to 5,000 mg/liter, depending largely on the sugar required in the process. Sugar-processing wastes range from 500 to 1,500 mg/liter 5-day B.O.D., while milk-product wastes and meat-packing wastes range up to 2,500 mg/liter. All the food-processing wastes are easily metabolized biologically. The rapidity with which they are metabolized is the source of their disposal problem.

Paper Wastes. The production of paper results in large volumes of soluble wood chemicals to be disposed of. The disposal of paper-mill wastes is probably the number one waste disposal problem in this country. The sulfite pulping operation produces wastes with a B.O.D. between 10,000 and 15,000 mg/liter. The prime organics are the wood sugars, which are biodegradable, and lignins, which are only slightly biodegradable. Paper coatings using starches and casein are used by some manufacturers and wind up in the waste waters. Wood fibers also make their way into the waste waters, as do large quantities of inorganics.

Textile Wastes. The textile-industry wastes can be divided into four groups: (1) cotton finishing, (2) wool scouring, (3) synthetic fibers, and (4) dyeing. Cotton-finishing wastes include kiering wastes, bleaching wastes, and desizing wastes. The cotton-kiering wastes result from washing cotton cloth after treatment with caustic soda. The kiering liquors have a pH of 11.5 and a 5-day B.O.D. of approximately 1,000 mg/liter. Bleaching is done with hydrogen peroxide or hypochlorite and does not yield a high B.O.D. waste. But the desizing waste is high in hydrolyzed starches. Combination of enzyme desizing and caustic kiering wastes at Hohokus, N.J., resulted in a 750 mg/liter 5-day B.O.D. waste. The peroxide bleach wastes contained less than 50 mg/liter 5-day B.O.D. At pH 11.5 the wastes are relatively stable. Lowering the pH to 9.0 to 9.5 permits the microorganisms to readily degrade the cotton-textile wastes.

The wool-scouring wastes are among the strong industrial wastes with a 5-day B.O.D. ranging from 1,000 mg/liter to 20,000 mg/liter. The wastes are high in fats and greases which are removed by the scouring process. The highly putrescible nature of wool-scouring wastes makes it difficult to dispose of satisfactorily.

The manufacture of synthetic fibers has created a new textile-waste disposal problem. Acetate rayon manufacture yields a waste high in acetic acid, which can be biologically degraded with ease. Orlon wastes contain sulfonated acrylonitrile, which can be biologically metabolized only after microbial adaptation. Other synthetic fiber wastes contain many unusual chemical structures, some easily metabolized and some not.

Dyeing of cloth is not an easy process and results in wastes containing acids and bases as well as dyes. One of the common acids used is acetic acid, which can be biologically degraded. Some dye molecules can be metabolized, while others cannot. The degradability of the particular dye is a function of its chemical structure; but so little information is available on biological degradation of dye molecules that it is not possible to state which ones can be metabolized and which ones cannot.

Petroleum Wastes. Petroleum wastes consist of oils, various hydrocarbons, phenols, and sulfides, as well as inorganics. The oils are removed largely by separators, but the soluble and emulsified materials can be readily metabolized biologically. At one time phenols were considered untreatable biologically, but it has since been shown that phenol concentrations of 2,000 mg/liter can be treated by the complete mixing activated sludge system.

Chemical Industry Wastes. It is impossible to discuss all the different types of wastes produced in the chemical industry. These wastes reflect the raw materials used, the intermediate products, and the final product. Most of the organic compounds produced by the organic chemical industry can be metabolized biologically, but many of them require adaptation before degradation is rapid. Some few compounds appear to be inert or highly resistant to degradation. Actually it is the resistance to biological degradation that is posing some of the major problems in the chemical industry. Instead of being destroyed by biological action the resistant compound persists for long periods of time in the watercourse and can create problems if the water is reused. A good example of this is with the tetrapropylene-based synthetic detergents. They are not degraded biologically to completion and persist with the production of foam. Nitro-substituted aromatics are another group of compounds not readily degraded but not completely resistant.

Many supposedly toxic or nondegradable organic compounds can be degraded if the microorganisms are allowed to adapt over a long period of time at a relatively low concentration. Formaldehyde, acetone, diethyl ether, and aniline were all organic compounds believed nontreatable by biological means, but they are completely treatable by activated sludge.

Five-day B.O.D. of Industrial Wastes. Unlike sewage, industrial wastes do not have a full complement of adapted microorganisms ready to go to work on the wastes, and often the wastes lack certain nutrients necessary for optimum microbial growth. This makes it difficult to determine the 5-day B.O.D. test of industrial wastes accurately. The sewage micro-organisms used as seed in the B.O.D. dilution water are not always adapted to all components of the waste either. The net result is a low 5-day B.O.D. Failure to run proper 5-day B.O.D. tests on industrial wastes has caused sanitary engineers to underestimate the strength of many industrial wastes.

In order to obtain proper B.O.D. values on industrial wastes, it is necessary to grow an adapted microbial seed which is ready to go to work on the wastes immediately. This is done best by maintaining a small laboratory activated sludge unit on the waste. The activated sludge unit can be either a quart jar or a gallon jug for simplicity. The activated sludge is maintained by a fill-and-draw operation, with one feeding of waste per day. The microorganisms can be kept in a rapid state of growth by wasting one-third of the activated sludge daily. The unit is fed and aerated for 23 hr; one-third of the mixture is wasted and the sludge allowed to settle for 1 hr before pouring off one-half of the remaining volume. The unit is made back to the original volume by the addition of raw wastes and aeration restarted. This procedure will yield an active microbial seed available at any time for B.O.D. work.

Treatability of Industrial Wastes. Often it is desired to know if an industrial waste is biologically treatable and to know the ultimate B.O.D. instead of the 5-day B.O.D. The procedure for obtaining an acclimated seed for the B.O.D. test will usually be sufficient to determine if a waste is easily treatable. Occasionally, a difficult waste will appear which will not yield an activated sludge. In order to determine the treatability of such a waste, the following procedure has been used successfully.

An activated sludge is built up on a readily degradable organic compound of similar chemical structure to the compounds in the waste. Sodium benzoate or phenol is used when the wastes contain aromatic compounds. Sodium acetate is used when the wastes contain simple hydrocarbons.

The activated sludge unit is operated on a once-a-day feed, fill-and-draw basis with no sludge wasting at an organic loading of 1,000 mg/liter. When the suspended solids reach 2,000 mg/liter, the easily degradable organic is replaced with the waste. By the use of the dichromate chemical oxygen demand test (C.O.D.) it is possible to determine if the organic matter is being degraded. If the C.O.D. does not decrease after a week's operation, the unit is restarted on the easily degradable organics. When the suspended solids have built back up to 2,000 mg/liter, 10 per cent of the easily degradable organics are removed from the feed and substituted by 10 per cent of the industrial wastes. The second day the ratio of waste to degradable organics is changed to 20:80. The third day the ratio is changed to 30:70. In ten days the unit should be converted to the full waste load. While this procedure seems long, it is often the only way to obtain an acclimated seed or unusual industrial wastes. A good example of this was in the degradation of sulfanilic acid. Sulfanilic acid is not easily degraded biologically but it is possible to slowly adapt a seed to metabolize this compound. A small laboratory activated sludge pilot unit is shown in Fig. 17-1.

Nutritional Deficiencies. Many industrial wastes do not form a balanced substrate for good microbial growth. When discharging industrial wastes into a watercourse, this nutritional deficiency can be an advantage since the rate of metabolism will be retarded. In biological waste treatment systems a nutritional deficiency can be a handicap. If a waste is to be stabilized biologically, it is essential that enough chemical elements be available for optimum cellular growth. The two most critical elements are nitrogen and phosphorus. But demineralized process waste waters can have other chemical deficiencies.

Fig. 17-1. Small plastic laboratory activated sludge unit for use in studying treatability of industrial wastes.

The disposal of industrial process wastes is the real challenge to the sanitary engineer. Unlike the disposal of domestic sewage, industrial waste disposal is not stereotyped and each new problem requires imagination and real engineering know-how.

Cooling Water

The third type of waste discharged from industries is cooling water. While it might seem that the relatively clean cooling waters would not be a waste problem, in many instances they are. The reason for this lies in the fact that cooling waters often create thermal pollution when discharged into receiving waters. Industries using large volumes of cooling water actually can raise the temperature of the receiving water quite materially. The biological effect on the receiving water is twofold: (1) decreasing the solubility of oxygen and (2) stimulating the biological activity. A power plant might add little organic pollution to a stream but it could create an oxygen-deficit problem by a thermal increase in a stream which already was partly loaded.

CHAPTER 18

Stream Pollution

Rivers and lakes are the natural result of the runoff of surface waters. As the waters flow over the ground surface, they pick up organic and inorganic materials which stimulate biological growth. A high mountain stream will show little biological life because of the lack of nutrients in the water. But a river flowing through a rich agricultural area will teem with biological life. To maintain a normal balance of biological forms in a river or a lake there must be a certain amount of organic matter as well as the necessary nutrient elements.

Streams have long been used for the disposal of domestic sewage and industrial wastes. As long as the load of wastes remains below the stream's assimilative capacity, the normal biological flora and fauna will be predominantly aerobic and beneficial to man. Once the load of wastes exceeds the stream's assimilative capacity, the higher forms of biological life are displaced by an excessive growth of bacteria and anaerobic conditions set in which are detrimental to man. It is the overload that establishes the normal pattern for stream pollution.

Normal Biological Cycle

The bacteria are the keys to the normal biological cycle. It is the role of the bacteria to convert the soluble organic matter into bacteria cells and inorganic elements. The inorganic elements are taken by the algae and converted into algae cells. The newly formed bacteria and algae become food for the protozoa, rotifers, and crustaceans. The animal forms and some of the large algae and bacteria become food for the minnows and tiny fish. The small fish become food for the large fish which become food for man. Man discharges his wastes back into the stream where the bacteria metabolize the organic matter and complete the cycle. Thus it is that waste discharge into streams is a normal part of the biological cycle (Fig. 18-1). Without the organic wastes there would be little fish for man to remove from streams.

Excess Wastes

While a certain amount of organic wastes are necessary for the normal biological cycle, too many wastes can destroy it. The key in the normal

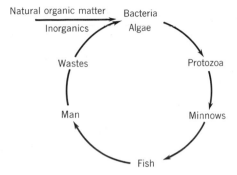

FIG. 18-1. Natural biological cycle in streams.

biological cycle is oxygen, which is essential for the development of the beneficial animal forms. As the organic waste concentrations increase, the bacteria growth is stimulated with a greater and greater demand for oxygen. As the bacteria decrease the oxygen level, the higher forms

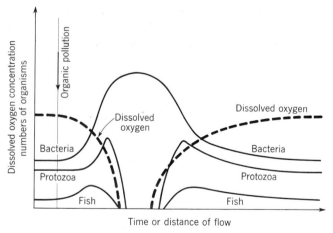

FIG. 18-2. Effect of pollution on biological life in a stream.

begin to die off. Sensitive game fish are the first to feel the effects of oxygen depletion, followed by the normal game fish and finally by the crustaceans, rotifers, and higher protozoa. The bacteria remain as the sole form of biological life. In the absence of dissolved oxygen the bacteria undergo anaerobic metabolism with its resultant odors and black colors. Even man is unable to live close to the polluted river. The effect of pollution in a stream is schematically illustrated in Fig. 18-2.

Oxygen Concentration

The primary problem in stream pollution lies in the low solubility of oxygen in water (Fig. 18-3). At 20°C only a little over 8 mg/liter of

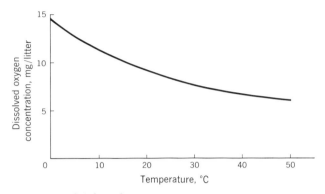

Fig. 18-3. Solubility of oxygen in water at various temperatures.

oxygen will exist in water. As the temperature rises, the oxygen solubility decreases and vice versa. This phenomenon is not good for the sanitary engineer because the rate of the biological activity and hence the rate of demand for oxygen increases with increased temperature and decreases with decreasing temperature.

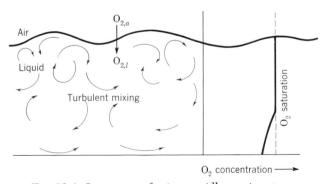

Fig. 18-4. Oxygen transfer in a rapidly moving stream.

In swiftly moving streams the oxygen used in biological metabolism is replenished quite rapidly. The water surface at the air-liquid interface is being continuously replaced by the turbulence of the stream movement. The oxygen-saturated layer is overturned and replaced by an oxygen-deficient layer. In this way the oxygen-saturated liquid is rapidly mixed throughout the stream to supply the microorganisms with sufficient oxygen (Fig. 18-4).

In a slow-moving stream, the surface water is not being turned over. The stream tends to flow in distinct layers. At the air-liquid interface the water is saturated with oxygen which can mix with the oxygen-deficient layers only by slow diffusion (Fig. 18-5). It is in the slow-moving streams

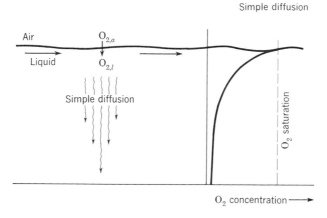

FIG. 18-5. Oxygen transfer in a slow-moving stream.

and in lakes that most of the stream pollution problems appear. The failure to replace the oxygen being used in the lower regions of the stream allows anaerobic conditions to develop.

Nutrient Concentration

The demand for oxygen in a stream is a direct result of the aerobic metabolism of the microorganisms in the stream. Normally, the metabolic activity in streams is limited by the lack of nutrients. The low organic level permits only a few bacteria and fungi to grow. These in turn limit the growth of the animals in the stream.

The sudden introduction of organic wastes increases the food concentration and allows the bacteria and fungi to speed up their rate of growth. The increased growth results in a proportional decrease in oxygen concentration in the stream. If the organic load is sufficiently high, the entire oxygen resources are depleted and the stream has become anaerobic.

Eventually, the rate of bacterial metabolism must decrease below the rate of oxygen transfer as the nutrient concentration becomes limiting. With aerobic conditions being reestablished, the predatory protozoa are stimulated by the tremendous bacteria population. The growth of the protozoa in turn stimulates the higher animal forms. As the oxygen level rises higher and higher, more and more complex animals can survive and find an increased food supply. Slowly, the original biological conditions are reestablished in the stream with a limited food supply and a low microbial population.

Biological Indicators

The microorganisms which exist in any body of water are direct indicators of the condition of that water. The types and numbers of microorganisms are all a direct result of the nutrient concentrations. It is impossible to have a high nutrient concentration and a low biological population in nature, or vice versa, unless there is a deficiency of a key element or a toxic condition.

In a low organic content stream the total number of microorganisms is low since food is limiting; but the number of different species of microorganisms is very high. The algae tend to grow better than the bacteria since they can utilize the inorganic elements in the water for their nutrients. The algae which die and lyse form most of the food for the bacteria, while the living algae are the major food sources for the protozoa and higher animals.

As we have already seen, the bacteria predominate completely in high organic streams under anaerobic conditions. A few fungi can survive near the air-liquid interface where a complete oxygen deficit does not exist. A few scavenger-type animals can survive in the bottom muds near the edges of the stream where a small amount of oxygen is available. The number of bacteria present under these conditions usually exceeds 10^7 or 10^8 per milliliter with a few well-adapted species. Thus it is that a few species of a large number of microorganisms characterize the polluted stream, while a few microorganisms of a large number of species characterize the clean stream.

Since the types of microorganisms in any body of water are a direct function of the environment and the nutrients, direct microscopic examination of the water can be used to determine the extent of pollution without any chemical tests. The results of the microscopic examination reflect the instantaneous condition of the water and are far more reliable than some chemical tests.

Use of biological indicators has never had much favor in sanitary engineering. The reason for this lies in the fact that the sanitary microbiologists have never presented their information in a simple enough form for the sanitary engineer to understand and to use. In the past the sanitary microbiologists have been interested in complete identification of every microbial form in a sample of water. This has resulted in long lists of microorganisms which were never the same from one stream to another. Efforts were made to correlate the chemical data with the microbial data, but actually there was never a consistent pattern of correlation. Quantitative enumeration of the microorganisms yielded no correlatable data. Needless to say, the sanitary engineer came to the erroneous conclusion that there was no correlation between the microorganisms and the chemical data. There is a definite, direct correlation between the microorganisms and the chemical changes which occur in any stream. There must be a direct correlation since the microorganisms produce the chemical changes.

A single quantitative enumeration of the general types of microorganisms such as bacteria, algae, fungi, protozoa, and rotifers is merely a qualitative measure of the stream condition. It has a limited value in itself. Two samples must be taken with a definite time differential be-

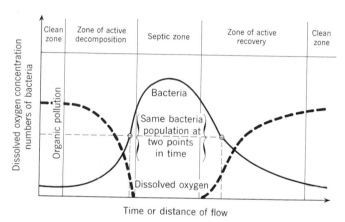

FIG. 18-6. Oxygen sag curve and corresponding bacteria population in polluted stream.

tween them to establish the rate of change of microorganisms. The microbial changes occurring at the onset of pollution are the same as for the recovery from pollution. The microbial population at 1 mg/liter of dissolved oxygen with 25 per cent organic stabilization is almost identical with the microbial population at 1 mg/liter of dissolved oxygen with 75 per cent organic stabilization (Fig. 18-6). The microbial characteristics are the same but the chemical characteristics of the stream are quite different. It is necessary to take two biological readings with respect to time to find out if the microbial changes are from 25 to 30 per cent stabilization or 75 to 80 per cent. In the 25 to 30 per cent stabilization zone the bacterial population is rapidly increasing, the free-swimming ciliates are increasing as are the flagellated protozoa. At the 75 to 80 per cent stabilization zone the bacteria, the free-swimming ciliates, are increasing (Fig. 18-7). These differences seem minor at first but upon repeated microscopic examinations even an untrained observer can spot the differences.

Oxygen Depletion

If nuisance conditions are to be avoided in a stream or if certain types of fish are to be maintained, it is essential that the dissolved oxygen level in the stream does not drop below certain critical values. It is the job of the sanitary engineer to determine the effect that a given waste will have on the oxygen resources of the receiving stream so that minimum dissolved oxygen can be maintained at all times. Since stream flow is never constant, the engineer must have a thorough understanding of hydrology to fix the waste load on any given stream.

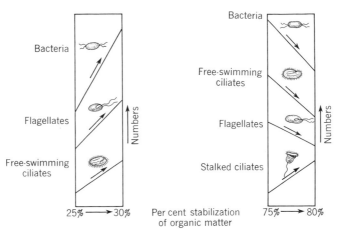

FIG. 18-7. Relative rates of microbial change at two levels of organic stabilization.

The most important aspect of stream pollution is the rate at which oxygen is depleted which is a direct function of microbial activity. There are two basic forms of microbial growths which affect stream conditions: (1) the dispersed growths and (2) the attached growths. The dispersed growths flow with the stream and act in the same manner as the microorganisms in a B.O.D. bottle. The initial food-microorganism ratio stimulates growth which results in rapid removal of oxygen and a reduction of organic matter in the stream. The rate of metabolism decreases as the $F:M$ ratio changes. Since the stream is flowing, the metabolic curve as indicated by oxygen depletion becomes stretched along the river in which distance is a function of time. If the stream is flowing rapidly, the distance of travel becomes quite great and vice versa. Thus it is with dispersed microorganisms that the oxygen-depletion curve will move up and down the stream with changes in flow.

The attached microorganisms include those growing on fixed objects such as stones or twigs and those deposited in the bottom muds and silts. These microorganisms obtain their organic matter from the water flowing past them. Normally, the rate of metabolism of the attached microorganisms is constant since the concentration of organic matter around them is relatively uniform and the space for attachment is limited. This means that the oxygen depletion curve remains relatively fixed rather than varying with stream flow. The attached growths can fluctuate somewhat with stream flow in those areas where all the surfaces for attachment are not full. This normally occurs in the latter ends of the recovery zone.

The dispersed microorganisms predominate in most stream pollution problems. Only in shallow streams do the fixed microorganisms exert a prime influence. Needless to say, there could be a good debate on the exact definition of "shallow streams." The engineer must recognize that the two groups of microorganisms exist and that most streams show a combination of both factors.

Reaeration

The depletion of oxygen from the stream results in a deficit at the air-liquid interface, causing oxygen from the air to enter the stream. The rate at which the oxygen enters the stream is a direct function of the dissolved oxygen deficit. The greater the deficit, the more rapid will be the rate of reaeration. The driving force in reaeration is the oxygen deficit at the air-liquid interface and not the oxygen deficit in the main body of the stream. A stagnant stream can have a low oxygen level in the lower depths of the stream and a low rate of reaeration. The oxygen transferred into the liquid must be transported to the lower layers of the stream and the liquid at the air-liquid interface must be maintained with a maximum deficit if reaeration is to be a maximum. This requires mixing or turbulence so that the high oxygen liquid at the surface is continuously replaced with oxygen-deficient liquid.

Since the physical characteristics of a stream are relatively fixed, the reaeration characteristics are also relatively fixed. The shallow, rapidly moving, turbulent stream will have a high rate of reaeration and hence will be able to absorb large quantities of organic pollution without creating nuisances. A deep, rapidly moving stream will have a lower rate of reaeration but will still be able to absorb relatively large quantities of organic matter. The slow-moving stream will have a low rate of reaeration. With a narrow, deep, stagnant stream, reaeration can be ignored for all practical purposes.

The one physical factor which must always be carried in the back of the mind is temperature. As the temperature increases, the rate of microbial activity increases, but the solubility of oxygen in water decreases. Thus it is that summer operations which often include minimum stream flows form the basis of investigation and analyses for the maximum quantity of organic pollution which can be safely discharged to the stream.

Oxygen Sag

Considerable effort has been put into the mathematical consideration of oxygen depletion. The most common form of the oxygen-sag equation is given below:

$$D = \frac{k_1 La}{k_2 - k_1} (10^{-k_1 t} - 10^{-k_2 t}) + Da(10^{-k_2 t}) \qquad (18\text{-}1)$$

where D = oxygen deficit (mg/l)
Da = initial oxygen deficit (mg/l)
La = ultimate B.O.D. (mg/l)
t = time (days)
k_1 = deoxygenation constant
k_2 = reaeration constant

This differential equation has a series of constants which must be carefully evaluated. The k_1 term reflects the rate at which the bacteria demand oxygen and is calculated from the B.O.D. test, by running 1-, 3-, 5-, and 7-day B.O.D. determinations. The k_2 term is the reaeration characteristic of the stream and varies from stretch to stretch in most streams. While k_1 can often be determined in the laboratory, k_2 can only be determined from field studies. The complexity of the oxygen-sag equation and the lack of accuracy in determining k values which can be used in the field have limited the use of this equation. Yet it remains the only method for approximating the oxygen sag before pollution has been added to the stream.

Where a stream is receiving some pollution below maximum, it is possible to use Churchill's empirical method for the direct determination of a particular stream's characteristics. Churchill postulated that the depletion of oxygen between two relatively close points on a stream was a direct function of the organic load, the temperature, and the stream flow. He expressed his equation as follows:

$$\text{D.O. drop} = A(\text{B.O.D.}) + B(\text{temp.}) + C(\text{flow}) + D \qquad (18\text{-}2)$$

The constants A, B, C, and D are determined from field data. It has been found from a large number of field measurements that A, B, C, and D are constant for fairly wide variations in B.O.D., temperature, and flow. The field data permit the determination of the minimum point in the oxygen-sag curve so that the D.O. drop between the point of pollution introduction and the minimum point on the oxygen-sag curve can be determined directly. Once the constants are determined, the maximum B.O.D. load can be calculated for maximum temperature and minimum flow. The direct method has the advantage of compensating for the actual stream characteristics. The disadvantages lie in the fact that the assumption is made that the stream conditions remain constant over the ranges of temperature, flow, and load studied. Where stream data are available, the direct method of Churchill has the advantage of speed in calculating the ultimate load a stream can assimilate. Where data are not available, the differential form of the oxygen-sag curve must be used to estimate the ultimate load.

The mathematical solutions to the oxygen-sag curve are of value to the engineer as long as he is aware of the fact that the results are no better than the assumptions he makes in evaluating the various constants. The oxygen-sag curve is determined by the microorganisms and not by mathematics. If conditions change so that the microorganisms change their rate of metabolism, the mathematical results will not be correct. The mathematics is of value only between uniform stretches of the river, where metabolism is uniform. Once again the engineer must temper his equations with an understanding of the microbial phenomena.

CHAPTER 19

Theory of Biological Waste Treatment

Fundamental microbiology offers the means for the sanitary engineer to base the design of biological waste treatment systems. It is important for the engineer to realize that all microbial systems operate on the same general biochemical principles and that the differences between the various biological systems lie in the environment imposed by the mechanical aspects of the system.

Microorganisms can continuously remove organic matter from liquid wastes by only one method, synthesis into new protoplasm. It is possible for the microorganisms to absorb large quantities of organic matter onto their cell surfaces but unless this absorbed organic matter is assimilated into protoplasm the rate of absorption will approach zero. Since a definite quantity of organic matter is required to form the building blocks for the microbial cells and a definite quantity of organic matter must be oxidized to form the energy necessary for synthesis, a relationship exists between the removal of organic matter and the cells synthesized, together with the oxygen consumed.

$$F = K_1 S + K_2 O_s \tag{19-1}$$

where F = organic matter removed, mg/liter ultimate B.O.D.
S = synthesis, mg/liter volatile solids
O_s = oxygen for synthesis, mg/liter oxygen uptake
K_1 = 1.43
K_2 = 1.0

The two constants are used to convert the units of measurement to milligrams per liter of oxygen.

It can easily be seen how Eq. (19-1) can be applied to aerobic systems, but anaerobic systems pose a different problem. There is not a direct oxygen uptake in anaerobic systems, only indirect oxidation. But it is oxidation, anaerobic oxidation, with the utilization of organic matter being related to the quantity of oxidation. Because of the problems created by these two basic processes, each will be studied separately in order to prevent confusion.

Aerobic Metabolism

The aerobic systems are the primary systems for waste treatment and include activated sludge, trickling filters, oxidation ponds, and composting. The basic equation is expressed directly in terms of dissolved oxygen uptake as long as the systems remain truly aerobic. Since there is a definite quantity of energy required to produce a definite quantity of protoplasm, we can set up an equation between synthesis and energy.

$$O_s = 0.70S \qquad (19\text{-}2)$$

Substituting this into Eq. (19-1), we have an equation which expresses the removal of organic matter in terms of synthesis.

$$F = 2.13S \qquad (19\text{-}3)$$

The problem remaining is to determine exactly what is synthesis. It is not a direct measurement of the increase in mass but rather is the sum of the increase in active mass plus the decrease in active mass due to endogenous metabolism.

$$S = Ma + K_3Mat \qquad (19\text{-}4)$$

where Ma = active microbial mass, mg/liter volatile solids
   t = time, hr
   $K_3 = 0.006$
Substituting this into Eq. (19-3), we have the following:

$$F = 2.13\Delta\,Ma + 0.0129Mat \qquad (19\text{-}5)$$

This is the basic equation which can be used in the design of aerobic waste treatment systems. Care must be taken with this equation, as it is not a single direct equation that can be used by itself. It must be combined with other equations in order to be solved. The most important item missing as far as design is concerned is the rate of the reaction. Unfortunately, there has not been sufficient fundamental research in this phase of sanitary engineering to evaluate all the rate factors.

There are two basic rate patterns in any biological system, an increasing rate and a decreasing rate. The increasing rate of reaction occurs during the log growth phase, while the decreasing rate of reaction occurs in the declining growth phase as well as the endogenous phase.

Log Growth. Sanitary engineers have tried unsuccessfully to utilize log growth for the stabilization of wastes. During log growth the rate of synthesis, and hence the rate of stabilization, is a maximum. This means simply that the most wastes can be stabilized in the shortest period of time. Unfortunately, the microorganisms have two biochemical factors which have prevented use of log growth in waste stabilization. In order to remain in log growth the instantaneous food to microorganism ratio

($F:M$) must always be above 2.5. With domestic sewage this would mean that the active mass could not exceed 150 mg/liter. The only time that a domestic sewage, aerobic biological treatment system has such a high $F:M$ ratio is during the initial startup. With such a high food level the microorganisms will not form floc, but will remain dispersed throughout the liquid. The effluent from the log growth phase then is high in unstabilized organic matter and dispersed bacteria. It is only in a few industrial waste treatment systems that log growth is in any way significant.

Declining Growth. Declining growth starts at the end of log growth when the concentration of organic matter becomes the limiting factor in the growth of new cells. The rate of metabolism in the declining growth phase cannot be determined, since it is not constant but is continuously changing as the organic concentration decreases. The declining growth phase ceases when the instantaneous $F:M$ ratio has dropped to 0.006. Most biological treatment systems are operated between declining growth and endogenous so that the terminal $F:M$ ratio can be used to determine the final organic concentration remaining for the active mass present. It should be stressed that all the organic matter has essentially been stabilized, i.e., removed from solution and converted into cellular protoplasm, by the end of the declining growth phase.

Endogenous Phase. Growth does not cease in the endogenous phase but is exceeded by the rate of cellular degradation so that there is a decrease in active mass in the endogenous phase. The small amount of synthesis which occurs in the endogenous phase results in a small removal of organic matter but at a very slow rate. The rate of reaction in the endogenous phase has been determined fairly accurately. Data to date indicate that the active mass decomposes at the rate of 0.6 per cent per hour. Ma breaks down into two components, one of which is oxidized and the other is inert. The oxidized portion decomposes at 0.5 per cent per hour, while the inert portion is formed at 0.1 per cent per hour. Converting the oxidized mass into oxygen utilization, we find that the rate of oxygen utilization is 0.7 per cent per hour.

$$O_e = 0.007Mat \qquad (19\text{-}6)$$

The endogenous oxygen uptake is the only method available to date for the engineer to determine the active mass in any given biological sludge. This method consists in allowing the microorganisms to stabilize the excess organic matter and then determining the rate of oxygen uptake either polarographically or on a Warburg. Figure 19-1 shows a curve for oxygen uptake of an activated sludge which contains some unstabilized organic matter. The tapering off of the rate of oxygen uptake signals the completion of stabilization and the beginning of endogenous metabolism. Thus, it is possible to determine the active mass Ma and the inactive mass Mi of any biological mass.

$$\text{Total mass} = Ma + Mi \qquad (19\text{-}7)$$

Normally, the volatile solids determination of the suspended microbial mass yields a measure of the total organics in the mass. From the endogenous oxygen uptake data Ma is determined and Mi is then calculated.

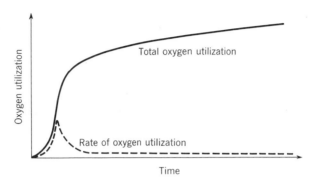

FIG. 19-1. Typical Warburg oxygen-utilization curves.

Anaerobic Metabolism

Essentially there is very little difference between the basic equations for anaerobic metabolism and aerobic metabolism. The quantity of organic matter to form a unit of protoplasm is the same in both systems. The metabolic patterns have many of the same basic structures. Needless to say, the energy required to produce a unit of protoplasm must also be the same in both systems. The only difference between the two systems lies in the energy mechanisms and energy yields per unit of organic matter metabolized.

The anaerobic reactions must be considered in two stages: (1) acid formation and (2) methane formation. In both of these reactions the only oxygen which can be added must come from water. All the remaining oxygen must come from the organic matter being decomposed. Acid formation results from the degradation of carbohydrates and proteins. Carbohydrates generally decompose into volatile acids, formic, acetic, and propionic. Under certain circumstances butyric and valeric acids are also formed. For practical purposes it can be assumed that the carbohydrates break down into a mole each of formic acid, acetic acid, and propionic acid per mole of glucose, or for simplicity, 3 moles of acetic acid. The formation of each mole of acetic acid is equivalent to 0.14 mole of oxygen. Translating this into milligrams per liter we find that 1 mg/liter of acetic acid is equivalent to 0.07 mg/liter of oxygen.

As a comparison between the aerobic and the anaerobic metabolisms let us assume that 100 mg/liter of glucose is synthesized into cells. In an aerobic system there would be 50 mg/liter of volatile solids formed and 36 mg/liter of oxygen utilized if we neglect endogenous respiration. Under anaerobic conditions there would be 9 mg/liter of volatile solids formed and 88 mg/liter of acetic acid produced. In the aerobic system the sludge formed is approximately $5\frac{1}{2}$ times that formed under anaerobic conditions. The 88 mg/liter of acetic acid has 94 mg/liter of oxygen equivalent still unstabilized.

The anaerobic degradation of proteins results in the formation of amino acids which are hydrolytically deaminated to form the equivalent hydroxy acid. The hydroxy acid is in turn reduced to the saturated acid as well as being oxidized to formic acid plus a saturated acid with one less carbon than the initial amino acid. All the amino acids do not form volatile acids so that volatile acids cannot be used as a direct measure of degradation when working with proteins. Instead, ammonia is used as the guide to the energy change. Each mole of ammonia released is equivalent to 0.5 mole of oxygen; 1 mg/liter of ammonia nitrogen is equal to 1.1 mg/liter of oxygen.

Methane formation occurs from further metabolism of the volatile acids formed from the acid degradation of the carbohydrates and the proteins as well as from degradation of the fatty acids. The methane bacteria obtain the equivalent of 0.13 mole of oxygen in the oxidation of acetate to methane and carbon dioxide. One mg/liter of methane is equivalent to 0.26 mg/liter of oxygen.

It can be seen that anaerobic metabolism is not efficient in producing cellular protoplasm, but then the production of cellular protoplasm is not the goal in biological waste treatment systems. The typical anaerobic waste treatment system is the sum of the acid formation and the methane formation. Since methane is soluble only to the extent of 17 mg/liter, complete anaerobic metabolism would produce an effluent low in organic matter and there would be a minimum of sludge for disposal. While we shall see that the current emphasis is on the aerobic treatment systems, there is little doubt that anaerobic treatment holds promise for the future.

CHAPTER 20

Trickling Filters

The trickling filter is the most widely used aerobic biological waste treatment system. Its simplicity of design and operation has ensured its popularity with engineers. In the past 50 years the trickling filter has undergone definite structural changes but the basic process has remained constant.

Initially, sewage filters were merely an extension of sand filters used in water purification. The organic-laden liquid was allowed to pass slowly through coarse sand where microorganisms removed the organic matter. The need for greater loadings per unit volume of filter prompted the use of gravel, small stones, and finally rock. The future trend indicates that plastics will replace rock in the filter.

Description

Trickling filters have been adequately described as merely a pile of rocks over which sewage or organic wastes slowly trickle. A schematic cross section of a trickling filter is shown in Fig. 20-1. The sewage is

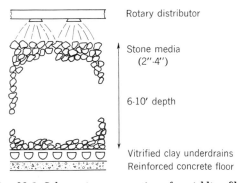

FIG. 20-1. Schematic cross section of a trickling filter.

introduced onto the filter by a rotary distributor which is driven either by electric motor or by hydraulic impulse. Where hydraulic head is available, the latter method is always used. In the early days fixed nozzle distributors were used on filters, but these have all but been replaced.

199

The rotary distributor is so designed that the wastes are discharged at a uniform volume per unit of filter surface. The wastes flow by gravity over the stones and into an underdrain system. All the liquid is collected into a main effluent channel which flows to a final sedimentation tank.

The depth of rock in the filter varies from a minimum of 3 ft to a maximum of 10 to 14 ft, with an average stone depth of 6 ft. The size of the rock ranges from 1 to 4 in. in diameter, and the rocks are carefully selected for their spherical shape. Flat stones tend to compact too tightly and reduce the essential void volumes. The choice of filter-stone size will be dependent upon the waste characteristics. With a high organic load per unit volume of filter it is essential to have large stones if the biological growth is not to fill all the voids and clog the filter. The relationship between organic loading, biological growth, and filter stones will be discussed in detail in a later section.

The underdrain system is normally vitrified-clay-block construction.

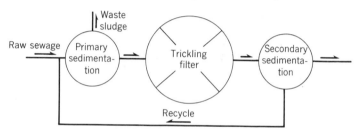

FIG. 20-2. Schematic diagram of a single-stage trickling filter.

Vitrified clay has a high structural strength and is able to withstand both the abrasive and corrosive action of the wastes. In effect the underdrain system is a network of tiny channels which rapidly discharge their flow into the main collection channel which normally runs along the middle of the filter. An equally important function of the underdrain system is to allow sufficient air to circulate through the filter. The underdrain channels must be designed so that they can carry the maximum quantity of air and liquid to permit the filter to operate at its maximum efficiency. Too-small channels will result in flooding, thereby blocking the air flow and allowing the filter to go anaerobic.

The final sedimentation tank is an integral part of the trickling filter and must be considered as such. The function of the sedimentation tank is to remove the large masses of biological growths which have dropped from the filter stones. The removal of these settleable biological growths accounts for 10 to 20 per cent of the efficiency of the filter in removing organic matter. The schematic diagram of the trickling filter systems is shown in Fig. 20-2.

Microorganisms

Although classed as an aerobic treatment device, the trickling filter is not a true aerobic device but rather must be considered as a facultative system. As the filter begins operation, it is primarily aerobic. The build-up of microorganisms creates an anaerobic layer at the stone interface which is as essential to the filter operation as the aerobic activity which occurs at the upper microbial surface.

The microorganisms in the trickling filter reflect the facultative nature of the filter. The predominant microorganisms are bacteria: aerobic, facultative, and anaerobic. The obligate aerobic sporeformers, *Bacillus*, are easily found in the upper, aerobic surfaces. On the other extreme the obligate anaerobe, *Desulfovibrio*, can be found at the microbial stone interface where oxygen is completely void. The great majority of bacteria in the filter are facultative, living aerobically as long as dissolved oxygen is present and anaerobically when the oxygen is depleted. The facultative bacteria include varied species of the genera *Pseudomonas*, *Alcaligenes*, *Flavobacterium*, *Micrococcus*, as well as members of the family of the Enterobacteriaceae.

Fungi are also present in trickling filters. Being aerobic organisms, the fungi live only in those regions where dissolved oxygen exists. The fungi must compete with the bacteria for their food and are not overly successful. The competition for food limits the growth of the fungi under normal environmental conditions. With unusual industrial wastes or at low pH levels the fungi can predominate over the bacteria in the filter, but this is the exception rather than the rule.

The upper surface of the filter can support algae. The algae do not contribute to the over-all stabilization of the wastes since they live on the excess inorganic ions in the wastes. The need for sunlight by the algae for energy definitely limits their growth to the surface of the filter, but their growth can be great enough to bring about clogging of the filter surface.

In addition to the microbial plants, there is an ample animal population in the filter. The protozoa are the predominant animals with all forms from the Phytomastigophora to Suctoria. The Phytomastigophora exist in the upper layers where the organic concentration is high enough to allow them to compete with the bacteria. Ciliata can be found all the way through the filter in the aerobic sections. The free-swimming ciliates predominate at the filter surface, while the stalked ciliates predominate in the lower regions. The protozoa predomination will vary from filter to filter and even within a single filter in response to the changing food supply and the environmental conditions.

Higher animals include worms, snails, and insect larvae. These animals feed on the microorganisms in the filter and live in the upper aerobic areas. For the most part they contribute little to the action of the filter other than removal of some of the lower forms of microorganisms. Figure 20-3 shows the biological growths on the surfaces of the filter stones.

FIG. 20-3. Filter stones coated with microbial growths.

Theory of Filtration

A close examination of the filter (Fig. 20-4) shows that the liquid is distributed uniformly over the surface of the stones from the rotary distributor. The liquid forms a thin layer as it flows over the surface of the top layer of stones and passes onto lower layers of stones. Since the rotary distributor does not dose a single point on the filter surface continuously, there will be a definite time lag between dosings. The liquid will pass over the stones as a surging wave very quickly and leave a thin coating of liquid on the stones (Fig. 20-5). The thin coating of liquid will absorb oxygen from the air in the voids and permit the microorganisms to carry out aerobic metabolism.

Oxygen Absorption. Oxygen is absorbed from the air in direct proportion to the deficit existing in the liquid at the air-liquid interface. The partial pressure of oxygen in the air in the void spaces is the driving force causing the oxygen to dissolve in the liquid. If the liquid contains less oxygen than saturation, there will be a definite drive to reestablish equilibrium at saturation. The rate of oxygen transfer will be a function of the liquid characteristics and the oxygen deficit. With most wastes the liquid characteristics are fixed and can be expressed as a constant K.

$$\frac{dO_2}{dt} = K[O_2(sat.) - O_2(actual)] \tag{20-1}$$

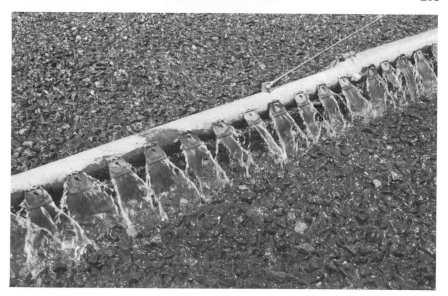

Fig. 20-4. Distribution of sewage onto filter stones from rotary distributer.

It is readily apparent that the maximum rate of oxygen transfer will occur when the oxygen concentration in the liquid is zero.

It must be realized that the interfacial liquid layer will become saturated with oxygen and that the rate of oxygen transfer will fall to zero unless the oxygen in the interfacial layer is transferred into the lower areas where an oxygen deficit exists. One method of oxygen transfer is by simple diffusion from the saturated surface layer to the unsaturated lower layer. Diffusion is a slow process and is governed by the concentration gradients. The best method for oxygen transfer is by the creation of turbulence. Turbulence causes the replacement of the oxygen-saturated surface layer by an unsaturated lower layer. The process of turbulence resulting in changing of liquid layers at the air-liquid interface is commonly referred to as *surface renewal*. Surface renewal is the key to oxygen transfer and hence the upper limit of microbial metabolism.

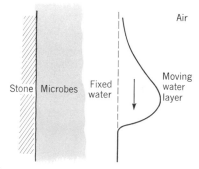

Fig. 20-5. Schematic diagram of liquid passing over surface of trickling-filter stone.

The total quantity of oxygen which can be transferred into a unit volume of the filter per unit of time is primarily related to the area of biological growth exposed to a continuously moving air supply. Indirectly the exposed growth surfaces are related to the surface area of the stone media. A 1-in. stone filter has approximately 37 sq ft of surface area per cubic foot, while a 4-in. filter has only 9.5 sq ft of surface area per cubic foot. From a purely oxygen transfer standpoint, the 1-in. filter would have a greater rate of oxygen transfer potential than the 4-in. filter. Yet, from a practical viewpoint, considering the volume of biological growths on top of the stones and the voids required for satisfactory air movement, the 4-in. filter could have a higher rate of oxygen transfer than the 1-in. filter.

Organic Removal. The incoming wastes contain a definite quantity of organic matter per unit volume which the filter must remove. As the liquid flows over the stones, it actually flows over the top of the microbial layer rather than through the microbial layer. The flowing liquid mixes with the bound water on the microbial surface. If the organic concentration in the bound-water layer is lower than in the flowing wastes, organic matter will be transferred from the flowing wastes to the bound-water layer. If the organic concentration in the bound-water layer is higher than in the wastes, the flow of organic matter will be in the opposite direction, with the flowing wastes increasing in organic concentration. In effect, the flowing liquid transfers its organic load to the bound-water layer by dilution with the most rapid transfer occurring when the incoming organic concentration is high and the bound-water concentration is a minimum. The concentration gradient concept is illustrated in Fig. 20-6.

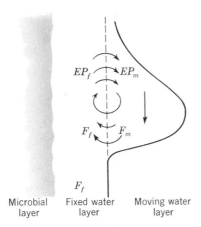

Fig. 20-6. Schematic diagram of transfer of materials between the moving liquid layer and the fixed liquid layer.

The filter will continue to remove organic matter only as long as the microorganisms reduce the organic concentration in the bound-water layer between applications of wastes. The top layer of microorganisms are stimulated to active metabolism by the high organic concentration as the fresh wastes flow past. A portion of the organic matter is converted into new cells, while an equivalent portion is oxidized to furnish the energy for synthesis. Although the organic concentration in the bound-water layer is high, the total quantity of organic matter per unit mass of microorganisms is low so that there is a rapid reduction of the organic matter before the next surge of wastes passes over the stones. Since the upper layers of the filter receive the highest concentration of organic matter, it is not surprising to observe that the maximum rate of microbial growth occurs at the surface and decreases through the filter.

The rate of microbial activity is directly proportional to the organic concentration. Thus it is that the rate of removal of organic matter by the microorganisms falls off with time, as shown in Fig. 20-7. It is impossible for the microorganisms to metabolize completely all the organic matter in the bound-water layer between dosing cycles. This means simply that it is impossible for a trickling filter to give 100 per cent removal of the organic matter washed out of the filter. The shorter the contact period, the shallower the filter, and the greater the hydraulic flow, the greater will be the organic concentration discharged from the filter. It has been said that trickling filters have a "nonremovable" fraction of B.O.D. that

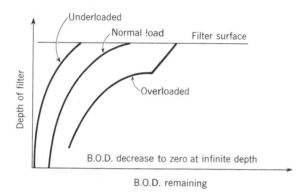

FIG. 20-7. Removal of B.O.D. at various depths of the trickling filter.

cannot be taken out. Actually, the B.O.D. that comes out of the filter can be removed as easily as that going into the filter under the proper conditions. Complete B.O.D. removal would result only if the filter was infinitely deep which is impossible from a practical standpoint.

The removal of organic matter by the filter is essentially a function of the microorganism present, the organic concentration applied, the rock size or microbial surface area, the time of retention of the liquid in the filter, and the temperature. The organic concentration applied to the filter is related to the waste itself and dilution by effluent recirculation. A high organic concentration is easier to remove than a low organic concentration, but a high organic concentration stimulates microbial activity so much that all the oxygen is depleted and the filter becomes anaerobic. In order to keep the filter aerobic, a high organic waste is usually diluted by recirculation to a concentration below that creating anaerobic conditions.

The time of liquid retention in the filter is a function of the microbial surface area and the hydraulic loading rate. At high hydraulic loading rates the time of liquid retention becomes so short that the microorganisms do not have time to stabilize the organic matter removed and soon reduce its removal. It can be seen that a high recirculation ratio decreases the organic concentration while shortening the liquid retention time. Both of these factors are detrimental to the removal of organic matter from the wastes.

The microbial surface area which actually comes in contact with the organic matter is a function of the filter media size and the depth of the filter. For a filter with a given stone characteristic, the microbial surface area can be related directly to depth. Thus, it is not surprising to note that the mathematical formulations for the removal of organic matter through a filter have been related to filter depth. It is important that the engineer realizes that depth per se is not important but only as it is related to the microbial surface area.

Bacterial Metabolism. It has already been indicated that the continuous removal of organic matter from the wastes being treated is related to the bacterial metabolism. As the organic matter is metabolized by the bacteria, a portion of it is oxidized for energy. In order to obtain the maximum energy yield, the bacteria must have dissolved oxygen. Thus, the rate of metabolism is limited either by the available organic matter or by the available oxygen. At low organic concentrations the organic matter controls the rate of metabolism.

$$\frac{dM}{dt} = kF \tag{20-2}$$

where F is the food concentration at time t. As the organic concentration increases, the rate of metabolism increases until oxygen becomes the limiting factor. The rate of metabolism is then controlled by the rate of oxygen transfer into the microorganisms. At very high organic loadings the rate of organic stabilization approaches a constant value. It can be seen that the filter has a definite limit as to the maximum rate of organic stabilization. Most filter loadings are set so that the rate of metabolism is not limited by the rate of oxygen transfer.

Temperature. The microbial surface area in a trickling filter is fixed so that the active microbial population in the filter is essentially constant. The microbial layer is very sensitive to temperature, increasing the rate of metabolism with rising temperatures and decreasing the metabolic rate with decreasing temperatures. Since the efficiency of organic removal is geared to metabolism, the filter's efficiency will definitely fall off in cold periods and increase in warm periods. The engineer must realize that the microbial activity in the filter is a function of liquid temperature and not air temperature. The air temperature does not have as strong an effect on the liquid temperature as might be expected from a casual examination (Fig. 20-8). In very cold weather the air temperature tends to cool the liquid but the microbial metabolism yields heat which tends to warm the liquid. The liquid temperature from the filter is then related to the quantity of cold air passing through the filter and the quantity of organic matter being metabolized. A heavily loaded filter will tend to remain at a higher temperature in the wintertime than a weakly loaded filter.

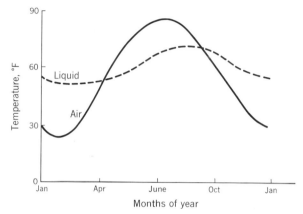

FIG. 20-8. Variation of air and liquid temperatures in a trickling filter during a typical year.

Microbial Mass. The microbial mass adheres to the rock surface as a result of the Van der Waal forces of attraction for two surfaces without sufficient repelling forces to cause them to stay apart. There is no special "glue" holding the microorganisms to the rock surface, only physico-chemical forces. The stone surfaces allow the microbial population to expand in only one direction, out from the rock surface. This means that the microorganisms build one on top of the other. As the microbial layer increases, the lower microbial layer does not receive any food other than in large particles adsorbed to microbial surfaces but not hydrolyzed before the growth covered them. The lower microbial layer does not obtain sufficient oxygen, as the upper microbial layer utilizes it all. The net result is that the lower microbial layer shifts to an anaerobic endogenous metabolism (Fig. 20-9). The end products of anaerobic metabolism are organic acids, aldehydes, and alcohols which slowly diffuse out of the cells toward the surface layer of microorganisms. Before long, the upper layer of microorganisms is receiving organic matter from both the top and the bottom. The net effect is to reduce the removal of organic matter from the incoming wastes and to cut the efficiency of the filter. Fortunately, the rate of diffusion of the anaerobic end products is slow and the quantity of end products is limited so that the effect on filter efficiency is a slow change which can be noticeable over a long period. As the cells continue endogenous metabolism, a point is reached where the cells die and lyse. The lysed products diffuse to the surface the same as the anaerobic end products. Lysing of the cell also destroys the surface responsible for holding the microbial mass to the stone surface. The microbial growth drops from the stone and the filter cycle starts anew.

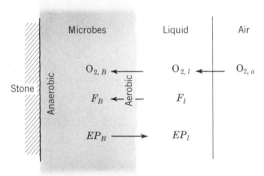

Fig. 20-9. Schematic representation of the microbial activity of a trickling filter. (O_2 = oxygen; F = food; EP = end products.)

The efficiency of the trickling filter is a minimum when the growth drops from the rocks and increases to a maximum when the stones are covered with a thin layer of microorganisms. As the layer of microorganisms continues to build up, the efficiency begins to fall slightly and then rapidly until the growth drops off. A filter which drops its excess growth continuously will not show the wide variations in efficiency.

Low-rate Trickling Filters

The low-rate trickling filter is the workhorse of small waste treatment plants requiring a high efficiency with a minimum of operations. The definition of low-rate filters varies depending upon the design criteria being considered. Hydraulically, a loading of 2 to 4 million gallons per acre per day (mgad), and organically, 10 to 20 lb 5-day B.O.D. per 1,000 cu ft of stone, define a low-rate trickling filter. With normal operation the low-rate filter will average 85 per cent B.O.D. reduction. By the addition of recirculation during periods of low flows so that the filter is always wet, it is possible to increase the filter efficiency to 90 per cent and even higher in some instances.

The low-rate filters are usually built with a depth of 6 ft. Yet, only the top 2 or 3 ft have any appreciable microbial growth. The active bacteria are in the top foot of the filter, while it is not surprising to find only a very thin film of growth on the bottom stones. The low-rate filter is in a sense overdesigned. The demand for oxygen is very low compared with the transfer ability so that the effluent from a low-rate filter always contains several milligrams per liter of dissolved oxygen. The excess oxygen and the conversion of all the excess nitrogen to ammonia permit the autotrophic nitrifying bacteria to grow in the lower portion of the filter. The autotrophic nitrifying bacteria oxidize the ammonia to nitrite and then to nitrate. Most low-rate filters show the excess nitrogen in the form of nitrates, indicating a very stable effluent.

High-rate Trickling Filters

The need for additional organic removal led to increased loadings on filters. It was found that if hydraulic loadings of domestic sewage were increased to 10 mgad, the filters were able to remove more pounds of B.O.D. per unit volume per day but the organic concentration in the effluent was very high. Recirculation of the filter effluent to produce the high hydraulic loading reduced the influent organic concentration and produced a lower organic concentration in the effluent.

Additional data showed that high-rate trickling filters could operate satisfactorily at hydraulic loadings between 10 to 40 mgad with organic loadings up to 90 lb/1,000 cu ft filter volume. Although a high-rate trickling filter can stabilize a large number of pounds of organic matter per unit volume, its efficiency for removal of organic matter is only 65 to 75 per cent. For strong wastes a satisfactory effluent can be obtained only with considerable recirculation. Mathematically speaking, an 8:1 recirculation is the upper limit. Above an 8:1 recirculation the influent concentration is only slightly above the effluent concentration and further recirculation becomes merely wasted motion.

The original high-rate filters were only 3 ft deep, since most of the microbial activity occurred in the upper 3 ft. Single-stage filters gave way to two-stage filters to obtain satisfactory effluent quality. It was realized that two 3-ft filters were not too different than a single 6-ft filter and so the high-rate filter increased its depth to 6 ft. The fact that the organic matter is distributed over a greater surface area prevents the development of the nitrifying bacteria and usually prevents the build-up of excess dissolved oxygen in the effluent. The high-rate filter is designed and operated very closely to the upper limits of microbial metabolism.

Intermediate-rate Trickling Filters

The hydraulic range between 4 to 10 mgad used to be considered a forbidden range for trickling filters. All sorts of troubles and ills were considered to lie within this range. Research has made some progress in finding out what the problems are within this area. Actually, there are many trickling filters operating in this hydraulic range without any problems; yet the problems have not been imaginary ones. It appears that the organic loading in the intermediate range stimulates considerable filter growth, while the hydraulic loading is not sufficient to push the excess growth from the stones. The net result is the tendency for the filter to clog. The clogging problem does not appear to be as severe if relatively large filter media are used, 2- to 4-in. stones.

Super-rate Trickling Filters

Recently, experimental filters on hydraulic loadings of 100 mgad and higher have been built. Instead of stone media these super-rate filters have used plastic media. Dow-Pac is the trade name for the plastic media manufactured by Dow Chemical Company. Their results indicate that super-rate filters can be built vertically rather than horizontally. Experimental filters have operated quite successfully at depths of 40 ft. The super-rate filters are a far cry from the original concept of trickling filters. Instead of depending upon adsorbed growths for organic stabilization, much of the microbial growth is suspended in the recycled effluent in the same manner as activated sludge. In essence, the super-rate filters are mechanically aerated activated sludge systems with a small portion of the growth adsorbed to the tile media. At organic loadings of 100 lb B.O.D. per 1,000 cu ft filter volume per day, efficiencies of 97 per cent have been obtained.

Modified Trickling Filters

There have been a large number of modified trickling filters proposed during the years, most of which have not progressed beyond the experimental phase. The most successful of the modified filters was made of extruded tile media manufactured under the trade name of Red Wing Tile. The tile filter media were designed to treat wastes too concentrated for rock media. Its use has been confined largely to the area of manufacture, Minnesota, because of the cost of manufacture.

A second tile medium used for experimental purposes was an extruded tile block with an air gap at the bottom of each block. The purpose of the air gap was to permit better air circulation. The design of the clay block permitted the filter to be laid out on a flat slab without a complex underdrain system. The last air gap formed the runoff channel for the effluent. Tests at organic loads up to 100 lb B.O.D. per 1,000 cu ft volume per day showed that the tile filter was more efficient than a 2- to 3-in. stone filter.

Ingram proposed that the organic wastes be applied at various levels rather than at the top only. In a sense he was trying to distribute the load over the entire volume of the filter. While the laboratory filter worked better than the conventional filter, it has not become practical to construct such a filter in full scale.

Design of Trickling Filters

The current design procedure for trickling filters is based on field experience rather than on fundamental theory. The correlation of operating data from many hundreds of trickling filters has yielded mathematical relationships which are used in design practice. The basic design criteria include the organic loading per unit volume of filter per day and the hydraulic loading with recirculation.

Thus far, the theoretical aspects of trickling filters have not progressed to the point where the engineer can design a filter for a given waste and predict in advance the results to be expected. The number of variables and their interrelationships in trickling-filter design have complicated the development of a complete fundamental theory. The development of a sound fundamental theory for trickling filters is one of the great uncharted areas of research.

Our present state of understanding of trickling-filter biochemistry can assist the engineer in making better evaluations of current design criteria. The following factors should always be in the back of the mind of any engineer when designing a trickling filter.

1. The microbial mass is essentially a function of the stone surface and should be kept as thin as possible.

2. The removal of organic matter by the filter is first by simple dilution which must be followed by microbial metabolism with a definite fraction being converted to new cells and a definite fraction being oxidized.

3. The microbial surface has a fixed rate of metabolism dependent upon the organic concentration up to the point where oxygen transfer becomes limiting.

Fundamentally, it can be seen that the filter should have as small a media as possible to yield the maximum surface area. The only problem is that as the size of the media decreases, the size of the void spaces decreases. The voids are essential for the microorganisms to fit, for the liquid to flow, and for the air to flow. If the filter is to remove the maximum pounds of organic matter per unit volume, there will be a very high rate of microbial synthesis. This means that the filter voids must be large enough to take the rapid growth and that the hydraulic load must be sufficient to keep the excess growth washed out of the filter. The high hydraulic load requires a large void space to keep the filter from flooding and to allow air to pass through the filter at all times. The high hydraulic load reduces the time of contact of the organic matter with the microorganisms in the filter. Both of these factors reduce the removal of organic matter.

A high organic concentration stimulates rapid synthesis which means large volumes of excess sludge for disposal. A low organic concentration leads to endogenous metabolism with a minimum of synthesis. The endogenous-type metabolism is more desirable than synthesis metabolism with regard to excess sludge disposal, but with a fixed microbial surface, the rate of metabolism of the endogenous system is approximately one-tenth that of the synthesis system.

The design of the trickling filter is a compromise between the major factors. In order to keep the biological growth as thin as possible, plastic or tile media will become more widely used as the economics permit. The hydraulic loadings will increase to keep shearing to a maximum and will tend to push the filter taller to retain a definite effluent quality. With increased loadings the secondary sedimentation tank will assume greater importance. But what the exact design criteria will be is still unknown for the lack of research. Until the fundamental research on filters is complete, the design criteria set forth in the Ten State Standards will be the primary design factors.

CHAPTER 21

Activated Sludge

The most versatile of the biological treatment processes is the activated sludge process. It can produce an effluent with any desired organic concentration from very high to very low. Recent developments in the field of industrial waste treatment have stimulated considerable interest in the activated sludge process and have added much to its rapid development.

The activated sludge process was developed in 1913 in England and remained unchanged for almost 30 years. When change came to the activated sludge process, it did not come from the design engineer or the research laboratory but from the operator. It is most intriguing to note that all changes to the basic process for over 40 years were due to the ingenuity of the operators in solving special problems. Only with advancing technology has the research laboratory begun to make its contributions to the modifications of activated sludge.

No other treatment process has more advantages or more disadvantages than the activated sludge process. The chief disadvantage with the activated sludge process lies in the lack of understanding of the basic process by both design engineers and plant operators. Activated sludge is a pure biological process and yet biology never entered into design or operation until the past few years.

Basic Process

Activated sludge is simplicity personified. It is formed merely by aerating a biologically degradable waste for a period of time until a large mass of settleable solids forms. The settleable solids are active masses of microorganisms and are designated as activated sludge. A schematic diagram of the basic process is shown in Fig. 21-1. The wastes enter the aeration tank after being mixed with return sludge. Diffused aeration along one side of the tank produces aeration and mixing as the wastes flow along the tank. The microorganisms aerobically stabilize the organic matter in the aeration tank and flow into a sedimentation tank. Sedimentation allows the activated sludge to flocculate and to settle out, producing a clear effluent of low organic content. A portion of the waste sludge is returned to the aeration tank as seed with the excess sludge being wasted to the digester either directly or through the primary tank.

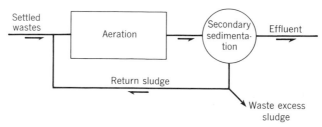

FIG. 21-1. Schematic diagram of a conventional activated sludge system.

Microorganisms

Activated sludge is made up of bacteria, fungi, protozoa, rotifers, and sometimes nematodes. The bacteria are the most important group of microorganisms, for they are the ones which are responsible for the stabilization of the organic matter and floc formation. All types of bacteria make up activated sludge (Fig. 21-2). The nature of the organic compounds in the wastes being stabilized determines which bacterial genera will predominate. A proteinaceous waste will favor *Alcaligenes, Flavobacterium,* and *Bacillus,* while a carbohydrate waste or a hydrocarbon waste tends toward *Pseudomonas* as well. At one time it was thought that activated sludge was formed by the bacterium, *Zoogloea ramigera* (Fig. 21-3), but it has since been shown that all bacteria contribute to the floc and that Z. *ramigera* probably plays a very minor role in activated sludge.

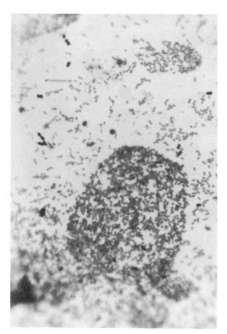

FIG. 21-2. Photomicrograph of the finger-like growths of *Zoogloea ramigera* sometimes found in activated sludge (4,000 ×).

Fungi are usually not desirable in activated sludge but are found under certain conditions. A high carbohydrate waste, unusual organic compounds, low pH, and nutritional deficiencies all stimulate fungi growths. Most of the fungi tend toward the filamentous forms which prevent good floc formation. With certain industrial wastes a nonfilamentous fungi, *Fusarium,* can produce an activated sludge with normal settling characteristics, but this is the unusual rather than the normal situation. Very little work has been done on identification of the fungi found in activated sludge, as mycologists have never considered waste treatment as a fertile field.

The protozoa do not contribute directly to the stabilization of the organic matter in the wastes being treated. The organic concentration is

FIG. 21-3. Photomicrograph of the fingerlike growths of *Zoogloea ramigera* (400 ×).

too low to support animal growth. But the protozoa can live off the bacteria which are utilizing the organic matter. This means that the Ciliata will be the primary protozoa of importance in activated sludge. With a high free-swimming bacteria population the free-swimming ciliated protozoa will predominate, which in turn results in stimulation of the Suctoria. As the bacterial population is reduced, the free-swimming ciliates give way to the stalked cilates (Fig. 21-4). The lower food level cannot support the high-energy-demanding, free-swimming ciliates.

Rotifers are not often seen in activated sludge systems. It has only been with the advent of the complete oxidation-type activated sludge systems that the rotifers have been seen as the predominant animal form (Fig. 21-5). The rotifers can utilize larger fragments of activated sludge floc than can the protozoa and survive after all the free-swimming bac-

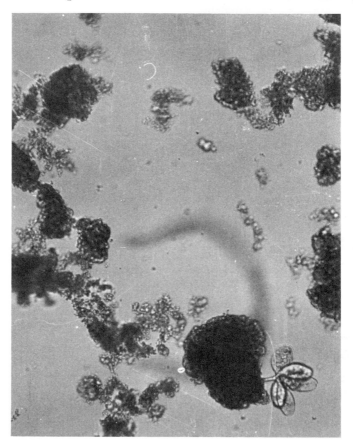

FIG. 21-4. Photomicrograph of normal activated sludge showing the stalked ciliate protozoa, *Epistylis* (400 ×).

teria have been eaten by the protozoa. The rotifers are indicators of an extremely stable biological system.

Fundamental Theory

Activated sludge is formed very simply by aerating the liquid wastes in the presence of bacteria until the bacteria have stabilized the organic matter. Activated sludge is not formed by special floc-forming bacteria, but rather is a normal phenomenon of all bacteria at a definite energy level. The best approach to the fundamental theory of activated sludge is to start with the raw wastes before activated sludge has developed and to follow it through each step.

Activated sludge can be formed from wastes high in colloidal solids such as domestic sewage or from a completely soluble waste such as an

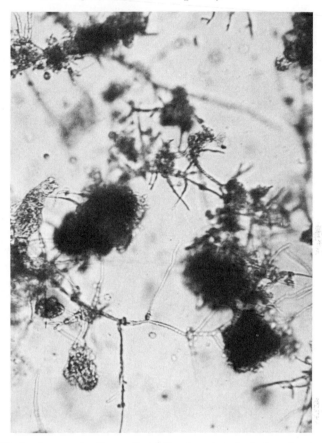

Fig. 21-5. Photomicrograph of a well-stabilized activated sludge showing a typical rotifer (400 ×).

industrial waste from the manufacture of synthetic chemicals. The formation of activated sludge is exactly the same in both extremes as long as the wastes are nutritionally stable, i.e., contain all the elements necessary for normal bacterial growth. Most wastes contain sufficient microorganisms to produce activated sludge without any seeding. These seed microorganisms come from soil contamination. In a few instances with industrial wastes it may be necessary to add some soil to give a suitable seed.

When aeration is started, the food-microorganism ratio is very large, so that the microorganisms are in an excess of food. The initial growth follows a log rate. As the bacteria begin to grow, the protozoa begin to grow. During the log growth rate the organic matter in the wastes is removed at its maximum rate with optimum conversion of organic matter into new cells. The energy level is sufficiently high to keep all the microorganisms completely dispersed. It is impossible to get activated sludge to form as long as the microorganisms remain in the log phase. The rapid rate of bacteria metabolism creates a very high oxygen demand. If aerobic conditions are not maintained by proper oxygen transfer, the rate of metabolism will not follow a log rate but rather an arithmetical rate until oxygen is no longer the limiting factor. The protozoa will be adversely affected if anaerobic conditions persist for very long.

The food-microorganism ratio drops rapidly as the food is consumed and new cells are produced. A point is reached where the food is no longer in excess but is the limiting factor in further growth. The growth phases have passed from the log growth to the declining growth. Further growth is directly proportional to the food remaining. Both the bacteria and protozoa begin to decline. A few of the cells begin to die and floc begins to form. In the turbulent aeration tank the bacteria are constantly being brought into contact with each other. As long as the bacteria have sufficient energy, the bacteria quickly split apart and continue their normal metabolic function. As the energy content of the system decreases, more and more bacteria lack the energy to overcome the forces of attraction between two cells once they have collided. The two cells move as a unit and soon become three and then four and so on until a small floc particle has formed.

The food concentration continues to drop and the microorganisms continue to increase but at an ever-slowing rate. The minimum $F:M$ ratio is reached at the end of the declining growth phase and the start of the endogenous phase. The $F:M$ ratio remains constant throughout the endogenous phase. It can be said that the system is essentially stable in the endogenous phase. Only a small quantity of food remains unmetabolized and it will be metabolized at a very slow rate. The demand for energy merely to stay alive is very low compared to that required for growth.

The bacteria are unable to obtain sufficient energy from the remaining food in the liquid around them and begin to metabolize food reserves within their own cells. The excess fats and carbohydrates are consumed first with the proteins being last. As the energy level drops, the rate of floc formation increases very rapidly. The free-swimming ciliated protozoa have a hard time finding enough bacteria to stay alive and they begin to die off. In the meantime the stalked ciliated protozoa begin to grow and they predominate during the early phase of endogenous metabolism. The bacteria begin to die off more and more. With death, an intracellular enzyme dissolves a portion of the cell wall allowing the nutrient contents left in the cell to diffuse out to furnish the remaining cells a little more food. The concept that the bacteria eat one another in a cannibalistic manner is absurd. Lysing allows the living bacteria to obtain nutrients from dead neighbors but only after the cell is dead. The only thing that remains of the bacteria is the nondegraded part of the cell wall and any nonlysed intracellular components. The major fraction of the cell wall remaining is believed to be the polysaccharide slime layer which the bacteria are unable to degrade. An enzyme capable of degrading the polysaccharide slime layer would be capable of degrading the cell wall since the slime layer is merely a waste product of part of the cell wall. For bacteria to make and hold such an enzyme would be like the proverbial container for a universal solvent.

If the aeration period was allowed to continue, the bacterial population would continue to decrease. The free-swimming ciliates would die out completely and the stalked ciliates would start to decrease, but rotifers would start to increase. The rotifers have the ability to eat small particles of the floc and do not depend upon the individual cells, as do protozoa. A very long aeration period would result in death of all biological forms and only the inert fraction of the cells would remain. Activated sludge is never allowed to go this far in its aeration cycle since it would take months to reach this point.

Normally, when the endogenous phase is reached, the tiny floc which has formed is separated from the major fraction of liquid by sedimentation. The concentrated floc is fed a fresh batch of organic matter. Since the quantity of microorganisms is higher than the first time around, the initial $F:M$ ratio is lower and the bacteria start out at a higher point in the growth cycle. A constant time period of aeration allows the system to progress a little further into the endogenous phase with each cycle. This results in better flocculation and a clearer effluent. Thus it is that the rate of organic removal is most rapid in the growth phase, while floc formation is best in the endogenous phase.

Theory of Operation

In the activated sludge system as shown in Fig. 21-1 the aeration tank has a fixed size, and hence a fixed retention period for any given flow. This means that the time for biological activity will be limited within the ranges of flow through the particular unit. If domestic sewage is considered, the aeration tank is designed for an average retention period and an average number of pounds of B.O.D. per unit tank volume per day. Within this average period the microorganisms undergo a continuously changing cycle along the growth curve, as shown in Fig. 21-6. When the return sludge is mixed with the raw wastes, the $F:M$ ratio determines how far back along the growth curve the microorganisms start. The time of retention in the aeration tank will determine how far along the growth curve the reaction will progress. With an organic loading of 20 to 30 lb B.O.D. per 1,000 cu ft aeration volume per day, and a 6- to 8-hr retention period, the microorganisms will be in the declining growth phase for approximately 2 hr and the remainder of the time in the endogenous phase. The long period of time in the endogenous phase permits the system to produce a well-clarified effluent of low organic content, but limits the system in its ability to handle shock loads.

One of the chief complaints against the activated sludge process is its apparent lack of stability. It is considered one of the hardest processes to operate, especially if the flow or organic loading varies widely. For

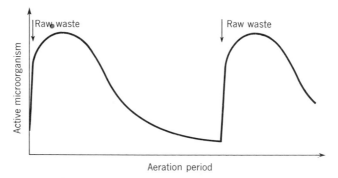

FIG. 21-6. Continuously changing growth curve during aeration period.

this reason activated sludge has been limited to the large cities where the variations in sewage strength and flows are damped out. This lack of stability in the conventional activated sludge system lies in the oscillating cycle that the microorganisms are constantly undergoing.

When the microorganisms reach the end of the aeration period, they are in equilibrium with the organic matter surrounding them. The number of living organisms is directly proportional to the organic matter available at this point. In normal activated sludge it is estimated that only 25 to 50 per cent of the volatile mixed liquor suspended solids are made up of living microorganisms. The remainder of the volatile solids is made up of inert organics left by dead cells. Returning the settled sludge to the head end of the aeration tank suddenly results in an increase in the $F:M$ ratio and stimulates rapid synthesis of new cells. The microorganisms metabolize the organic matter in an attempt to reestablish equilibrium.

In a normal cycle of domestic sewage, the flow and the organic strength are very low at night and increase during the day. This means that the microorganisms will extend much further into endogenous metabolism at night and will not be prepared to accept the sudden increase in organic load with reduced retention time in the morning. The net result is a decrease in process efficiency. As the day progresses, the time period of metabolism is decreased to a minimum. The tank size must be sufficient to allow the microorganisms to reach endogenous metabolism during the periods of maximum flow and organic loadings. If the aeration tank is not sufficiently large for this to occur, then the effluent will become turbid and the microorganisms will be lost in the effluent. The constant oscillation and continuous imbalance of the microorganisms are the basic problems in the successful design and operation of activated sludge systems.

Oxygen Requirements

The activated sludge process is an aerobic process, requiring an oxygen residual of at least 0.5 mg/liter at all times if the system is to op-

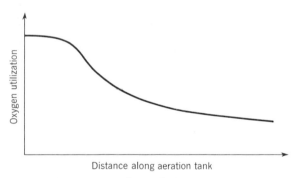

FIG. 21-7. Oxygen demand along aeration tank.

erate properly. Allowing the system to go anaerobic can result in damage to the protozoa which are needed for clarification and extends the time required for stabilization to occur. Oxygen is the primary limiting factor in conventional activated sludge systems. The demand for oxygen is a direct function of the biological metabolism. The greatest demand for oxygen occurs at the head end of the aeration tank when the food and microorganisms are mixed together. The oxygen uptake curve is shown in Fig. 21-7. If the demand for oxygen is greater than the supply, anaerobic conditions will set in and problems will develop in the operations. Low dissolved oxygen concentrations result in turbid effluents, since the protozoa do not develop to their fullest extent, and in filamentous bacterial growths which retard floc compaction and settling.

Oxygen is supplied to activated sludge by either mechanical or diffused aeration systems. Figure 21-8 shows a small diffused aeration plant. In the large installations diffused aeration employing large volumes of air at low pressure is the only method used. In domestic sewage systems it appears that 1 cfm of air per gallon of sewage at 5 to 6 psi will yield 5 per cent oxygen transfer efficiency. The low oxygen transfer efficiency of diffused aeration has long been considered one of the major engineering problems in activated sludge systems.

In diffused aeration the oxygen in air must be transferred from a bubble into the liquid and then into the microorganism. The fact that oxygen is a very insoluble gas is the chief factor against a high transfer efficiency. Oxygen transfer in activated sludge is very similar to that in trickling

Fig. 21-8. Small activated sludge aeration tank employing diffused aeration.

filters. The rate of oxygen transfer is a function of the oxygen gradient existing between the gas and the liquid, the surface area of contact between the liquid and the gas, the time of contact, temperature, and the characteristics of the liquid. The maximum rate of oxygen transfer for any given system occurs when the oxygen concentration in the liquid is zero. This gives the maximum driving force from the gas to the liquid. It is not surprising that the manufacturers of air-diffusion equipment rate their efficiency at zero dissolved oxygen since the efficiencies are a maximum at this point. Unfortunately very few activated sludge systems operate at zero dissolved oxygen so that the field equipment does not match the manufacturers' claims.

The surface area of contact between the gas and the liquid increases as the bubble size decreases, so that small bubbles transfer more oxygen than large bubbles. As the bubble size decreases, there is an increase in the resistance to gas interchange between the liquid and the gas. A point is reached where the size of bubble is so small that the rate of oxygen transfer decreases with further reduction in bubble size. The decreased oxygen transfer may well be due to reduced turbulence around the bubble as it moves through the liquid and the resultant failure to remove the saturated layer of oxygenated liquid from around the bubble.

The oxygen diffuses into the liquid from the bubble and saturates the surface film. Further oxygen transfer depends upon removal of the saturated oxygen film and its replacement by an oxygen-deficient film. Turbulence created during the bubble formation, its rise to the liquid surface, and its breaking at the liquid surface all assist in the surface renewal and oxygen transfer.

A rapidly rising bubble will be in contact with the liquid for a very short period and will not give as much oxygen transfer as a slowly rising bubble of the same size. The time of contact of the air bubble can be changed only by reducing the velocity of rise or the length of the rise. Too much reduction in the rate-of-rise velocity will cut down the mixing currents in the aeration tank and allow dead spots to build up.

Temperature reduces the solubility of oxygen in water but increases the rate of transfer because of its effect on the other characteristics affecting oxygen transfer. The chemical characteristics of the liquid under aeration affects the rate of oxygen transfer. A high salt concentration or organic concentration retards the rate of oxygen transfer. Sewage itself will retard oxygen transfer. One of the most controversial subjects today is the effect of synthetic detergents on oxygen transfer. There is no doubt that the presence of synthetic detergents in pure water has an effect on oxygen transfer, but there is reason to believe that the presence of synthetic detergents in sewage does not depress the rate of oxygen transfer any more than the other organic components of sewage. It is strange indeed with all the research on syndets and oxygen transfer that no one has ever undertaken a thorough study of sewage since all the data are aimed at sewage purification plants.

Thus it is that the efficiency of oxygen transfer occurring at the head end of the aeration tank where the sewage is first mixed with the activated sludge is lower than it will be after some of the organic matter has been metabolized. As the demand for oxygen decreases, the efficiency of oxygen transfer falls off. This means that the efficiency of oxygen transfer follows the function shown in Fig. 21-9. The gross efficiency of oxygen transfer is low in an activated sludge plant because of the poor transfer efficiency during the endogenous phase where there is little demand for oxygen. The problem of oxygen transfer, rate of stabilization, size of treatment plant, and costs are all interrelated factors which require compromise in order to obtain a satisfactory solution. A highly stable effluent cannot be obtained with a high-efficiency oxygen transfer system and a short aeration period with a high organic load. A high oxygen-transferring device such as a turbine aerator will produce poor results in a low oxygen-demanding system and will not be as economical as a low-efficiency diffused aeration system.

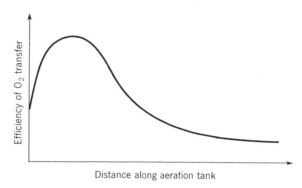

FIG. 21-9. Efficiency of oxygen transfer along aeration tank.

Nutritional Requirement

In all biological waste treatment systems, it is necessary that the microorganisms have all the necessary elements to form protoplasm. Domestic sewage contains all the elements that the bacteria require, but some industrial wastes are deficient in key elements. The primary nutritional elements missing in industrial wastes are nitrogen and phosphorus. The exact quantity of nitrogen and phosphorus required for a waste can be determined from the quantity of sludge produced per day. The pounds of nitrogen required per day will equal 10 per cent of the volatile solids, dry weight, produced each day, while the phosphorus requirements will be one-fifth the nitrogen requirements. The amount of nutrients which will have to be added daily can be determined from the quantity in the wastes and the quantity required.

The bacteria make use of nitrogen in the form of ammonia, but have the ability to utilize nitrites and nitrates as well as gaseous nitrogen under certain circumstances. Most facultative bacteria of importance in activated sludge have the ability to reduce nitrates to nitrites and through a series of intermediates to ammonia for incorporation into protoplasm. This assimilative reduction of nitrates and nitrites is specific for meeting their protoplasmic needs and can occur in the presence of excess oxygen, but will not occur as long as there is an excess of ammonia.

The use of gaseous nitrogen is limited to the *Azotobacter*, the nonsymbiotic nitrogen-fixing bacteria. In the absence of any nitrogen source

and in highly specific carbon sources it is possible to build up an activated sludge from *Azotobacter*. The *Azotobacter* fix only enough nitrogen from the atmosphere to meet their protoplasmic demands and do
not fix excess nitrogen for their neighbors, as is sometimes erroneously
thought. The nitrogen which appears in solution comes from lysing of
the *Azotobacter* and permits the normal bacteria to grow up in the solution also. There is no known record of nitrogen-fixing bacteria being
used in field installations, but they have been tried successfully in the
laboratory.

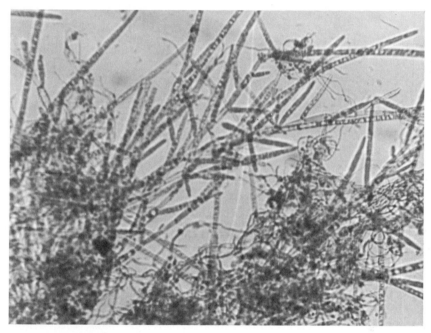

Fig. 21-10. Photomicrograph of activated sludge grown in a nitrogen-deficient environment, showing fungi growths (400 ×).

Other nutrient elements such as potassium, calcium, magnesium, molybdenum, cobalt, and iron are required in trace quantities. Normal
process contamination ensures that there will be sufficient trace elements
to meet the needs of the microorganisms.

One of the most important aspects of nutritionally deficient wastes is
their effect on biological predomination. A partially nitrogen-deficient
waste will stimulate fungi over the bacteria, since the fungi form protoplasm with a lower nitrogen content than bacteria (Fig. 21-10). The
fungi which predominate are filamentous and prevent good settling. If
the fungi had good settling characteristics, they would be as desirable

as bacteria since they can stabilize the organic matter like the bacteria. The same is true of a phosphorus deficiency.

Environmental Factors

The three environmental factors of importance in activated sludge systems are pH, temperature, and ORP. All three of these factors are intensity measurements rather than quantity measurements. We have already seen how pH affects biological growth. Between pH 6.5 and 9.0 we will see normal bacterial predomination and growth. Below pH 6.5 the fungi will begin to compete with the bacteria, with full predomination at pH 4.5. Above pH 9.0 we see retardation of the rate of metabolism. Thus it is important that pH be maintained at the proper level. The maintenance of a definite pH level is a function of the buffer capacity of the system. An activated sludge system must not only have the proper pH, but also sufficient buffer to resist a pH change. In most biological systems the bicarbonate buffer system predominates. In laboratory studies the phosphate buffer system is used in preference to the bicarbonate buffer since the pH can be held at a definite level over a wider range.

When an activated sludge system is metabolizing a neutral organic compound such as an alcohol, aldehyde, ketone, or carbohydrate, the pattern of metabolism results in the formation of acids which must be neutralized if the pH is to be maintained at the proper level. If sufficient buffer is lacking and the pH drops, the rate of metabolism will fall until the acids have been metabolized and the pH rises. There are many industrial waste systems operating over a longer aeration period than required, because of the lack of sufficient buffer to neutralize the metabolic acids. Unfortunately the measurement of pH at the start and at the end of the aeration period fails to show the effect of the lack of sufficient buffer. It is essential that the pH measurements be made at regular intervals along the aeration tank or aeration period to ascertain deviation from the normal.

The effect of temperature on biological reactions has already been discussed in detail. Increasing the temperature $10°C$ results in doubling the rate of biological reaction. With a high temperature the rate of reaction can easily exceed the ability of the system to remain aerobic. On the other hand a low temperature results in a slow rate of metabolism. The rate of metabolism can be increased by raising the microorganism concentration. Since the summer period is the most critical with regard to the streams' ability to absorb pollution, the treatment plant design is usually based to handle the average load at summer temperatures.

The oxidation reduction potential (ORP) is probably the most over-rated and misunderstood measurement in sewage purification. The ORP is a potential between the oxidants and the reductants without regard for the total quantity of each or their biological activity. This latter fact is most important since inert oxidized salts are measured as oxidants even though they have little effect on the biological reactions. A high ORP can result, even though the biological reaction is predominantly reducing. Actually, the ORP is not the measurement of the *cause* of the biological reaction, but rather is the resultant of the biological reaction. As the measurement of the resultants of the reaction, it is possible to correlate the ORP of any given system to its operating characteristics, but it is not possible to correlate ORP values directly between systems.

Process Modifications

The activated sludge process has not changed much since its inception in 1913, but there has been a series of minor modifications each designed to solve a specific operating problem. Activated sludge design has been one phase of sanitary engineering in which the operators have set the pattern for the consulting engineers to follow.

Tapered Aeration. The fact that the demand for oxygen decreased along the aeration tank led to the conclusion that the quantity of oxygen could be reduced at the end of the aeration tank. The number of air-diffuser tubes were increased at the head of the aeration tank and decreased at the end of the tank. Tapered aeration has become the standard method for arranging diffuser tubes in conventional activated sludge systems.

Kraus Process. Kraus in Peoria, Ill., was faced with the treatment of domestic sewage combined with a carbohydrate waste. The high carbohydrate concentration created a nitrogen and an oxygen deficiency. The net result was a light, poor settling activated sludge with a low efficiency of stabilization. Kraus solved his problem by aerating digester supernatant to form a highly nitrifying activated sludge with a high percentage of inert, readily settleable solids. The nitrifying activated sludge converted the high ammonia concentration in the digester supernatant to nitrates. The mixed liquor from the nitrifying process was discharged directly into the aeration tank with the raw wastes and the normal return sludge. The extra nitrogen from the digester supernatant supplied the bacterial demand for nitrogen. The nitrates also supplied the extra oxygen with the nitrogen being reduced to nitrogen gas and being blown out of the aeration tank. The extra inert solids from the digester supernatant added weight to the sludge and improved its settling characteristics. The schematic diagram of the Kraus modified activated sludge system is shown in Fig. 21-11. The Kraus modification has seen only limited use, since there are few plants faced with the same problems as the Peoria plant.

Step Aeration. New York City has made more contributions to the modification of the activated sludge process than any other city. The most important of these modifications was step aeration, in which the wastes were introduced into the aeration tank at several points rather

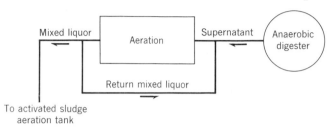

Fig. 21-11. Schematic diagram of the Kraus modification of activated sludge.

than all at once. Figure 21-12 shows a step aeration tank with four points of feeding. The splitting of the sewage flow reduced the F factor in the $F:M$ ratio and lowered the initial oxygen demand which had exceeded the oxygen supply when the system was operated as a conventional activated sludge plant. By having multiple feeds the demand for oxygen was spread over more of the tank and resulted in more efficient utilization of the oxygen. Step aeration is getting more and more use as

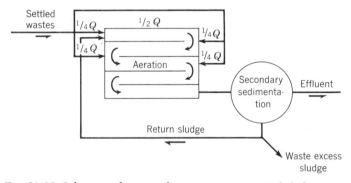

Fig. 21-12. Schematic diagram of step aeration activated sludge system.

a means of increasing the capacity of existing plants at a minimum of expense as well as being used in new construction.

High Rate. High-rate activated sludge is probably the most abused and misused of the activated sludge modifications. Although New York City cannot claim the origination of the high-rate modification, it can claim utilization of the process to its maximum efficiency. Basically, the activated sludge process is a highly efficient process, but there are times when there is a need for a less efficient process. Such was the case in

New York City. The activated sludge systems only needed to produce 50 to 60 per cent B.O.D. reduction with good removal of the settleable solids. The high-rate process filled the bill.

In high-rate activated sludge the $F:M$ ratio is kept high to give maximum synthesis by maintaining a low mixed liquor solids concentration M. While a conventional activated sludge plant carries 1,500 to 3,000 mg/liter MLSS, the high-rate activated sludge plant carries from 200 to 500 mg/liter. The high rate of synthesis gives a high rate of oxygen demand but this does not last long if F is not too high. In New York City it is possible to obtain 65 per cent B.O.D. reduction with only 2 to 4 hr aeration period. The short aeration period does not allow the microorganisms to move into the endogenous phase, but keeps them between the log growth phase and the declining growth phase. The failure of the microorganisms to form good floc results in loss of many of the microorganisms in the effluent. Efforts to translate the New York City data to other installations have met with some success and some failure.

Philadelphia's North Side plant suffered from its high-rate process because the demand for oxygen exceeded the system's ability to transfer it to the microorganisms. Their high-rate activated sludge was a typical filamentous sludge formed in a deficit of oxygen. The Los Angeles Hyperion plant failed to produce the desired results because of temperature and an increased rate of biological metabolism. New York City made high-rate activated sludge work because of a weak sewage and a low average temperature. Where similar conditions exist or where suitable oxygen-transferring devices are available, high-rate activated sludge can be operated with satisfactory results.

Biosorption. Biosorption is one of the few treatment modifications to claim two independent originators. Mansel Smith of Austin, Tex., and Wesley Eckenfelder of Manhattan College, New York, arived at Biosorption by separate pathways for different reasons. Mansel Smith was faced with the fact that it was impossible to expand the Austin Sewage Treatment Plant because of physical limitations, yet he had to take more sewage than the plant was designed for. Wesley Eckenfelder was looking for a process to absorb satisfactorily the shock loadings created by a cannery waste. Mansel Smith observed a phenomenon in sewage treatment that had been observed many times earlier but had always been chalked up as an experimental error. He found that if you mixed raw sewage and activated sludge together in an aeration vessel and removed samples at regular intervals for settling, the B.O.D. of the settled supernatant followed the curve in Fig. 21-13. He noted that there was an immediate drop in B.O.D., followed by a rise and then a second and final drop. The first drop had been noted by other investigators, but they had thought that the low point was an experimental error. Mansel Smith proved that the first drop was real and made use of the high absorptive properties of activated sludge.

The Biosorption process consists in mixing the raw sewage with a high concentration of activated sludge, 4,000 mg/liter, for 15 to 30 min to absorb all the colloidal organics. The sludge is settled for 1 hr and the concentrated sludge is aerated to stabilize the organic matter and to re-

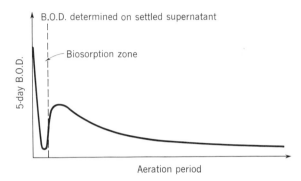

Fig. 21-13. Variation of 5-day B.O.D. of raw sewage–activated sludge mixture with aeration period.

new the activated sludge surfaces for further absorption. Figure 21-14 shows the basic Biosorption process. The process works best on colloidal wastes since the activated sludge cannot readily absorb large quantities of soluble organics in the short contact period.

At Austin, Tex., it was possible to convert the existing plant facilities over to Biosorption with excess facilities where a few months before the

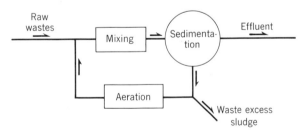

Fig. 21-14. Schematic diagram of the Biosorption activated sludge system.

same facilities had been overloaded. The same type of conversion was carried out at the Bergen County Sewage Treatment Plant, Little Ferry, N.J., where existing units were changed from an overloaded condition to an underloaded condition with surplus units.

Dispersed Aeration. One of the most extreme activated sludge processes was dispersed aeration in which the microorganisms remained completely dispersed without any flocculation. The absence of flocculation eliminated the need for secondary settling. Essentially, the system was so loaded that the microorganisms were always kept in the log growth phase. The rate of biological stabilization approached its maximum rate as did the rate of oxygen demand, but the total B.O.D. stabilization could never exceed 50 per cent. All the microorganisms produced by the stabilization of the wastes were discharged in the effluent, creating a very turbid effluent. Dispersed aeration has never advanced much beyond the laboratory stage, but it confirmed the fact that if the $F:M$ ratio is maintained at very high levels, the microorganisms will not floc but will be completely dispersed and the energy level of the system will be quite high at this nonflocculent stage.

Complete Oxidation. One of the most novel processes was found in research on dairy wastes by Hoover and Porges. They found that activated sludge rapidly removed the organic matter from solution and converted much of it into protoplasm which was degraded when all the organic matter was removed. This fundamental concept which was well known in the field of bacteriology had never been applied to a practical problem in waste treatment. They felt that it was possible for the sludge formed to be completely digested aerobically, thereby eliminating the need for anaerobic digestion facilities which were the bane of all small waste treatment plants. Rupert Kuntz at Pennsylvania State University was given the job of converting the concept of complete oxidation from the laboratory into full-scale plants. It was found that while it was impossible to burn up the activated sludge completely by aeration, the discharge of the small quantity of solids in the effluent did not create a problem in the receiving stream.

Complete Mixing. Small activated sludge plants were designed in such a manner that the raw wastes were always completely mixed with the aeration tank contents (Fig. 21-15). Although the small activated sludge plants operated with a minimum of technical supervision, little attention was paid to the fact that these small plants were not affected by the same factors which adversely affected large plants. The small plants were able to absorb satisfactorily widely fluctuating organic and hydraulic loadings. Because of a lack of interest of the consulting engineer in small activated sludge plants, the small plants were developed by equipment manufacturers and sold as a package. The package plants were usually sold directly to the contractor.

Following World War II with the expansion of industry and suburban development there was a definite need for a small sewage treatment plant with the result that there was stimulation of the package plant activity. Chicago Pump Company, which had one of the earliest activated sludge package plants, developed the Rated Aeration unit which was a

F<small>IG</small>. 21-15. Complete mixing activated sludge plant built with four modular units to treat cotton-textile wastes.

single-tank, complete oxidation system. Infilco Company converted their solids contact water treatment plant to an activated sludge plant under the title Aero Accelator. Yeoman's Brothers came out with the Cavitator, and Smith and Loveless had the Oxigest. All these processes operated on the basic principle of complete mixing and yet none of them were specifically designed around this principle. For the most part the basic design criteria of these package plants have not been divulged by the manufacturers. The Rated Aeration unit is known to be based on a 24-hr aeration period for wastes with less than 300 mg/liter 5-day B.O.D. Chicago Pump Company found that with this design criteria it was possible to produce an effluent with a low B.O.D. and little excess solids. Essentially, Chicago Pump Company's Rated Aeration was the field solution to the complete oxidation process as advanced by Porges and Kuntz. But there were definite operating problems which arose. The chief problem lay in the sedimentation section. Failure to return the activated sludge to the aeration tank resulted in anaerobic conditions. The microorganisms in the activated sludge reduced the nitrates which had been formed from prolonged aeration of the excess ammonia. The production of nitrogen gas caused the sludge to rise to the surface of the sedimentation section and be discharged in the effluent. The loss of the sludge was reduced by placing baffles parallel to the effluent weir just a few inches from the weir. Unfortunately, the sludge once risen, would not settle back and underwent anaerobic digestion with the creation of obnoxious odors. It was eventually necessary to abandon the combined aeration-sedimentation unit in favor of a separate sedimentation tank and positive sludge return to the aeration tank (Fig. 21-16).

The small package activated sludge plants represent the latest development in activated sludge. Like the process itself these package units are continuously evolving to find the optimum combination for most efficient operation. Strangely enough the design of these package units

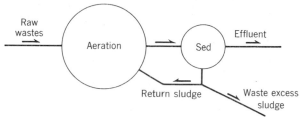

FIG. 21-16. Schematic diagram of the complete mixing activated sludge system.

and the various modifications of activated sludge have never been based on the biological fundamentals. While the modifications have solved biological problems, the results were obtained by trial and error and not with any fundamental concepts.

Biological Design of Activated Sludge

The design of any biological waste treatment system can be made properly only if the designer has a thorough understanding of the microbiology and the biochemistry of the process. Students of sanitary engineering are usually amazed to find that the engineers never consider microbiology in the design of waste treatment systems and that the sanitary bacteriologists are not interested in the design of treatment systems.

In the past 30 years there has been a tremendous amount of research on the fundamentals of the activated sludge process which the engineer has never made use of because he never understood them or because the researchers never translated their results into practical terms which the engineer could understand. This vast storehouse of information has lain dormant like the pirate treasure of old waiting for someone to uncover the key to its use in the field.

The purpose of activated sludge or any biological waste treatment process is to remove the organic matter as completely as possible from the raw wastes with the least expenditure of time and funds. The nature of the raw wastes and their flow characteristics are set and the treatment process must be designed around them. Only in a few industrial situations can the wastes be discharged to a surge tank for leveling out the organic strength and flow.

One of the basic problems with conventional activated sludge has been the variable predomination of the microorganisms. The organic matter in the raw wastes stimulates certain bacteria species in the aeration tank. As the organic matter is removed and the microorganisms pass into the endogenous phase, the primary group of microorganisms which were responsible for stabilizing the organic matter die off and lyse. A second group of bacteria utilize the lysed products of the primary group and predominate at the end of the aeration tank. When the sludge is returned to the head end of the aeration tank, the primary group of bacteria have to grow all over again. With a long aeration period the number of primary bacteria can drop off to a level that requires a long recovery period to handle sudden increases in waste flow or strength. This is very important in activated sludges treating strange industrial wastes, and it is one of the reasons activated sludge is slow in responding to shock loadings.

The only way to keep the bacteria species uniform is to keep the organic level uniform. This is impossible to do with a variable waste, but the variations can be kept to a minimum. The $F:M$ ratio is the key to bacterial growth. A high $F:M$ ratio has been shown to yield rapid growth while a low $F:M$ ratio depresses the apparent growth. In conventional activated sludge, both F and M are constantly changing, with the $F:M$ ratio going from a maximum to a minimum with each cycle. If the raw wastes were diluted with the entire contents of the aeration tank as in complete mixing, the aeration tank would become a large surge tank, damping out wide fluctuations in F. With complete mixing in the aeration tank, F becomes a minimum value with minimum fluctuations and the microbial activity at any point in the tank is the same as at any other point. The microbial variation over the normal growth cycle tends to approach a point function rather than a wide band as in conventional activated sludge. Thus it is that the complete mixing system offers the most advantages from a microbiological standpoint and the least disadvantages.

In any waste disposal problem the organic load is fixed so that the $F:M$ ratio can be varied only by adjusting M. It is possible to operate at any point on the bacterial growth curve by adjusting the $F:M$ ratio. It has already been pointed out that certain phases of the growth curve do not yield satisfactory operations as far as waste treatment are concerned. With complete mixing systems the operations will range at some point in the declining growth phase.

One of the most recent innovations has been the operation of complete mixing systems at the lower end of the declining growth phase where sludge increase is minimal. The excess sludge is allowed to be discharged with the effluent without causing nuisance conditions. Let us consider what would happen with conventional domestic sewage having an average 5-day B.O.D. of 200 mg/liter and volatile suspended solids of 240 mg/liter. It has been estimated that 73 mg/liter of the 240 mg/ liter of volatile suspended solids would be biologically inert organic matter and thus not metabolized by the bacteria. There would also be 80 mg/liter of inorganic suspended solids in the raw wastes, making a total of 153 mg/liter of inert suspended solids that must be discharged from the system. If the effluent carries all of the solids discharged from the system, then the 153 mg/liter of suspended solids must appear in the effluent once the system has reached equilibrium. Very few control authorities would permit the discharge of an effluent having this high suspended solids even if they were completely inert. Since the biological system could not discharge the inert solids without some active solids, there would be considerably more solids in the effluent than 153 mg/liter.

FIG. 21-17. Small activated sludge plant treating ice-cream wastes.

The ratio of active solids to inactive solids in the effluent would have to be the same as it is in the mixed liquor. If the sludge was 50 per cent active and 50 per cent inactive, then the effluent would contain over 300 mg/liter of suspended solids, or more than was actually in the incoming waste. It is doubtful if total oxidation will have much application to the disposal of domestic sewage unless the solids are removed at periodic intervals, as they cannot be discharged with the effluent.

On the other hand, a soluble industrial waste can be operated close to total oxidation with only a slow build-up of inert solids. The excess solids in the effluent would be in direct proportion to the organic matter being treated with approximately 10 to 13 per cent of the organic matter appearing in the effluent as inert solids.

In most instances activated sludge must have a means of solids separation and wasting. One of the major problems still to be solved with activated sludge is the efficient removal and concentration of the sludge from the liquid. Current methods depend upon gravity separation and are limited in their ability to concentrate the sludge. With 50 per cent sludge return and a good settling sludge, we can expect 3,300 mg/liter MLSS in the aeration tank, but no more. At 100 per cent sludge return we can get 5,000 mg/liter MLSS and at 200 per cent sludge return we can get 6,700 mg/liter MLSS, but economics has already called a halt to the practical limit of sludge return. In small systems it is possible to

Fig. 21-18. Industrial waste complete mixing activated sludge plant treating antibiotic wastes.

have 200 per cent sludge return, but large systems are hard pressed to justify 50 per cent sludge return. Thus we have upper limits of total mass, inactive as well as active. If the rate of sludge wasting is high compared to the total mass of sludge in the aeration tank, the majority of the total mass will be active mass; but if the rate of wasting is low, then the major fraction of the total mass will be inactive rather than active.

If the sludge is predominantly active, it will require further stabilization before it can be ultimately disposed of. Normally, anaerobic digestion is used for sludge stabilization but recently in small systems aerobic digestion has accomplished the same purpose. Once again economics determines which process should be used since both will produce a stable material which can be dewatered by filtration. If the sludge is predominantly inactive, there is no need for further stabilization and the sludge can be dewatered directly.

Microbial Indicators for Operation

Since the microorganisms follow definite patterns in all the activated sludge systems, it is possible to determine the operating characteristics of the various systems by making routine microscopic examinations of the activated sludges. The microorganisms act as instantaneous biochemical chemical indicators and eliminate the need for complex chemical tests, especially in small systems.

Activated sludge floc does not have a definite size or shape, but rather is a heterogeneous agglomeration of microorganisms. A good activated sludge will flocculate readily and produce an effluent that is free of dispersed bacteria. The absence of dispersed bacteria is usually accompanied by a relatively active population of stalked ciliated protozoa such as *Vorticella* and an occasional rotifer or the free-swimming ciliate, *Stylonichia*. Chemically, the 5-day B.O.D. test of the settled effluent will be between 5 and 10 mg/liter. If the sedimentation tank fails to yield good separation of the activated sludge, the 5-day B.O.D. test will be about equal to the weight of volatile solids lost.

While it is important to recognize a good activated sludge under the microscope, it is also important to recognize the symptoms of a poor activated sludge and what is responsible for the problem. A sudden loss of protozoa can result from anaerobic conditions or toxic materials. If the lack of oxygen has killed the protozoa, it will be noted that the number of free-swimming bacteria has suddenly jumped. With toxic materials there will usually be no sudden increase in free-swimming bacteria, but rather a decrease. An activated sludge fed a toxic material will often look perfect and will have a sparkling effluent for a short period.

The growth of filamentous microorganisms such as actinomycetes or fungi is readily observable under the microscope. The filamentous microorganisms keep the floc from being compact. Filamentous forms usually result in nutritional deficiencies such as nitrogen and phosphorus, low pH levels, and low oxygen levels. The most critical factor for normal activated sludge systems is the low oxygen level, while industrial wastes normally suffer from nutritional deficiencies. It appears that the filamentous microorganisms are able to compete with the bacteria at oxygen tension levels between 0 and 0.5 mg/liter. The bacteria begin to metabolize anaerobically, while the filamentous microorganisms continue with aerobic metabolism. The energy balance favors the filamentous forms in this region. If the system goes completely anaerobic, the bacteria will reestablish predominance since the filamentous microorganisms are generally strict aerobes.

Recently, it has been noted in the total oxidation systems that there is a tendency for filamentous microorganisms to predominate over a long period of operations unless sludge is wasted at reasonable intervals. It appears that the filamentous microorganisms are able to utilize slowly the inert polysaccharide material produced by the bacteria, thus giving the filamentous forms a source of food that is unavailable to the bacteria. Further research is needed on this point.

The best activated sludge has been produced in the total oxidation systems where the bacterial activity is so low that only a few rotifers are visible as living animals. The 5-day B.O.D. test of such a system is usually under 5 mg/liter.

As the organic load increases on the activated sludge, the number of bacteria in the system increases with stimulation of the free-swimming ciliates. A good activated sludge system with a B.O.D. between 10 mg/liter and 20 mg/liter will have a high population of free-swimming ciliates and a high population of stalked ciliates. The two forms will be in almost equal predomination. Increasing the organic load will stimulate the free-swimming ciliates to very high levels. At relatively high organic levels the tiny flagellates appear in competition with the bacteria. It will be noted that the dispersed bacteria population is very high. One key bacterial form is the spirillum. The presence of large numbers of spirillum is good evidence of a poor activated sludge system.

Recognition of key microbial operation characteristics can come only from repeated observations of various activated sludges. With experience it is possible to analyze any activated sludge system just from microscopic examination of the activated sludge and a routine knowledge of how the system is operated. The importance of this technique to the field engineer cannot be overemphasized.

CHAPTER 22

Oxidation Ponds

One of the newest treatment processes to obtain rather extensive use, especially in small municipalities, is the oxidation pond or lagoon (Fig. 22-1). Like the other biological treatment processes very little is known of the microbiology or biochemistry of the oxidation ponds. Construction has been on trial-and-error basis, with research only recently becoming of value in setting forth certain design criteria.

Description of Process

An oxidation pond is a large shallow pond in which wastes are added at a single point either at the middle of the pond or at one edge and the effluent removed at a single point at one edge. The ponds are usually 2 to 4 ft deep, although deep ponds, 10 to 20 ft, have been used quite successfully. The minimum depth is controlled by weed growth. It has been found that in most instances a depth of 2 ft is sufficient to prevent growth of weeds. The maximum depth is related in part to mixing or the lack of mixing, depending on which is wanted. The shallower the depth for a given waste, the greater will be the surface area. Since mixing is largely dependent upon wind currents a shallow pond with a large surface area will generally have better mixing than a deep pond with a small surface area.

Originally, oxidation ponds were used as secondary treatment only. The heavy solids were removed in primary sedimentation tanks prior to being digested in anaerobic digesters. The primary effluent was treated in the oxidation ponds. The cost of construction and operation of the sedimentation tanks and the digesters caused the construction of oxidation ponds to treat the entire sewage. Today, the tendency is for complete treatment in oxidation ponds.

Theory of Operation

One of the most confused items in biological waste treatment is the theory behind oxidation ponds. Very few engineers really understand how and why an oxidation pond works and what results can be obtained.

By its name the oxidation pond is an aerobic treatment device. It may have anaerobic zones in it so that it often becomes a combination aerobic-anaerobic device or a facultative device much in the same manner as the trickling filter.

The stabilization of the organic matter is brought about by the bacteria, although certain of the flagellated protozoa do give an assist. These bacteria will produce organic acids under anaerobic conditions or carbon dioxide and water under aerobic conditions. Since the B.O.D. of the effluent will be low only when carbon dioxide and water are the end products of metabolism, efforts are made to keep the system aerobic. It

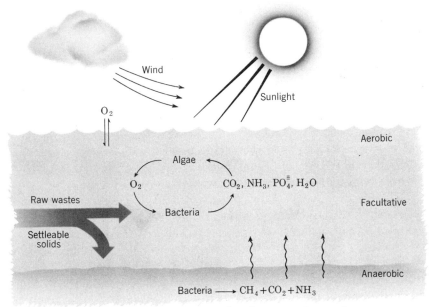

Fig. 22-1. Schematic diagram of oxidation pond operations.

has been thought that algae are responsible for keeping the oxidation pond aerobic, but this is not entirely true. Aeration is actually a combination of surface aeration and microbial reaction, with the former being as important if not more important than the latter.

It has been shown that algae take the energy from sunlight and the inorganics in water to synthesize their protoplasm. One of the side reactions in the synthesis of protoplasm is the release of oxygen. It has been felt by many engineers that the release of oxygen by the algae supplied the necessary oxygen for aerobic stabilization of the organic matter by the bacteria. Actually, the algae cannot produce enough oxygen to meet the demand of the bacteria and the protozoa unless they have an

additional source of nutrients than that from the wastes being metabolized by the bacteria.

If we examine the metabolic cycle closely (Fig. 22-1), we would see
that the bacteria metabolize the organic matter aerobically with the
production of bacterial protoplasm, carbon dioxide, and water as end
products. The algae take the carbon dioxide produced by the bacteria,
water, and inorganic minerals and convert them into algal protoplasm.
There is a release of oxygen in proportion to the carbon dioxide reduced.
The oxygen actually does not come from the carbon dioxide but rather
from water, which furnishes the source of hydrogen for the reduction. The
quantity of oxygen liberated is somewhat less than the quantity of oxygen
utilized by the bacteria since a portion of the oxygen utilized by the
bacteria goes into the oxidation of hydrogen. This means that an additional source of oxygen must be supplied from another source if the system is to remain aerobic.

In hard-water areas there are considerable bicarbonates and carbonates
in the water to furnish the algae an additional source of carbon not connected with the metabolism of the wastes. The utilization of the bicarbonates and carbonates can result in considerable excess of oxygen being
released by the algae than required by the bacteria. The water can even
become supersaturated with oxygen.

Close examination of a material balance of this cycle shows that the
bacteria will produce a maximum of 0.4 lb of organic matter from each
pound of ultimate B.O.D. stabilized. To keep the system aerobic from algal
metabolism there would be 0.3 lb of organic matter produced by the
algae. Since the raw sewage has approximately 0.7 lb of organic matter
per pound of ultimate B.O.D., the net effect is that 0.7 lb of organic matter in the form of sewage is converted to 0.7 lb of organic matter in the
form of algae and bacteria protoplasm. In essence, all that has happened
is conversion of the organic matter from one form to another form. In
the presence of sunlight the algae do not have a demand for oxygen since
the sun supplies their necessary energy but at night without sunlight the
algae would demand oxygen in the same manner as the bacteria for
endogenous metabolism. The oxygen demand of the protoplasm formed is
less than the oxygen demand by the cells forming the waste so that the
rate of oxygen demand has been reduced by this conversion.

Thus far, surface aeration has been ignored as a source of oxygen. It
cannot be ignored, since in the shallow ponds with large surface areas
the wind action across the pond surface will cause the breaking of the
surface and mixing of the pond contents. If the pond is deficient in oxygen, then surface aeration will tend to drive oxygen from the air into the
liquid to correct the oxygen deficit. At night when the algae are demand-

ing oxygen the same as the bacteria surface, aeration must supply the oxygen to keep the system aerobic (Fig. 22-2). Since the engineers have only recently recognized the fact that surface aeration is a factor in the oxidation pond, very little is known as to the rate of oxygen transfer possible under varying wind conditions.

The sizes of most oxidation ponds are designed sufficiently large that they act as sedimentation ponds as well as oxidation ponds. The inert solids which are not biologically metabolized and many microorganisms settle out in the pond and do not go out in the effluent. The microbial population is a maximum where the organic matter is introduced but falls off toward the effluent pipe. Many of the microorganisms undergo endogenous metabolism, die, and settle out before reaching the effluent pipe. In this manner the oxidizable organic matter is removed from the liquid before it is discharged into the receiving stream. It can be seen

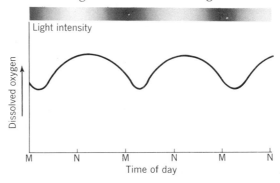

FIG. 22-2. Oxygen variations in an oxidation pond, according to the time of day.

that eventually the oxidation pond will have to be dredged or the walls made higher to compensate for the volume lost by sedimentation of inert matter in the raw wastes and from the microorganisms.

Microorganisms

The microorganisms in oxidation ponds are the same as those existing in the other treatment processes, with the bacteria and algae predominating and the protozoa and rotifers being present under certain loading conditions. The predomination of various species of microorganisms will depend upon the loading factors and the physical design of the oxidation pond. An oxidation pond at a low organic load will have a different microbial population than one at a high organic loading. A completely mixed oxidation pond will have a different group of microorganisms than will a pond having little mixing.

The predominant bacteria will be the *Pseudomonas, Flavobacterium,* and *Alcaligenes.* It has been said by some that the rapid die-off of coliform bacteria points to the production of antibiotics by the algae or other bacteria. Actually, there has been no evidence of antibiotic production in oxidation ponds, but there is a rapid die-off of coliforms. The reason for this lies in the competition of the predominating microorganisms for food and the high population of predatory protozoa. Coliforms have been shown to lose out in the competition for food with the bacteria in the above three genera in all waste treatment systems.

The algae predomination will depend upon the types and concentration of nutrients available. The phytoflagellates such as *Euglena* and *Chlorella* predominate in those areas where the nutrient level is quite high. These phytoflagellates can metabolize either photosynthetically or chemosynthetically and can actually assist in stabilizing a portion of the organic matter. Their size prevents major competition with the bacteria for the food. The small phytoflagellates require high energy and predominate over the other algae at the high nutrient levels. The filamentous green algae appear where the nutrient level drops off and the energy yield is not sufficient for large masses of the active phytoflagellates. Some of the more common forms are *Spirogyra, Vaucheria,* and *Ulothrix.* The over-all nutrient balance will determine exactly which forms will predominate.

The protozoa will show predomination in accordance with the nutrients present. The high organic concentration at the inlet will often stimulate the flagellates such as *Chilamonas* but they quickly give way to the free-swimming ciliates such as *Colpidium, Paramecium, Glaucoma,* and *Euplotes.* As the bacteria population decreases, the stalked ciliates such as *Vorticella* and *Epistylis* grow. If the organic load is very low and the oxygen level is sufficiently high, the higher animal forms such as *Daphnia* and *Rotaria* appear. The higher animals can utilize many of the algae as well as the bacteria and assist in producing an effluent low in algae as well as bacteria. Once again the animals are important in the clarification step rather than the purification step.

Design of Oxidation Ponds

Very little is known about the design criteria of oxidation ponds. Most of the ponds are designed on the basis of pounds of B.O.D. per acre per day. The common design criteria is 20 to 30 lb per acre per day, with some ponds reaching 50 lb per acre per day. The actual design criteria lie in the rate of oxygen transfer possible into the oxidation pond. If the organic matter were completely dispersed throughout the pond each day, the organic level in a 4-ft pond depth in terms of B.O.D. would be increased only 2 mg/liter at 20 lb per acre per day or 5 mg/liter at 50 lb per acre per day. At this low organic concentration an oxygen transfer rate of 0.12 mg/liter/hr from surface aeration could stabilize the entire organic load at 20 lb per acre per day and only 0.31 mg/liter/hr would handle the 50-pound load. Unfortunately, the organic matter is not completely dispersed throughout the oxidation pond and the demand for oxygen in some sections of the pond is greater than in others. Even with greater oxygen demands in sections of the ponds, they are kept aerobic, indicating that with complete mixing the B.O.D. load on the oxidation ponds could be increased over their present loads.

At an organic loading of 50 lb B.O.D. per acre per day with raw sewage a 4-ft oxidation pond would have an average retention period of 43 days. This large retention period and the shallow depth would eliminate any hydraulic flow pattern due to the addition of the wastes. Flow is governed by the wind currents. For this reason it has been American practice to construct the oxidation ponds circular, with the influent in the center and the effluent at one edge (Fig. 22-3). Distribution is just as good in

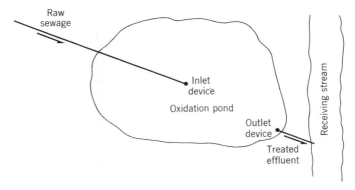

Fig. 22-3. Schematic layout of typical oxidation pond.

this type of setup as an elaborate inlet structure and elaborate outlet structure. Operations may be better in the center inlet if prevailing winds change considerably.

The question of a single pond versus series ponds has caused quite a flurry of differences of opinion. The purpose of the series ponds is to effect better circulation of the wastes into the pond. At the current loadings, 20 to 30 lb B.O.D. per acre per day, there has been no advantage to series ponds over the single pond. The advantage of the series ponds comes only when the organic load is raised considerably above current criteria. It should be possible to construct a series of ponds with recirculation around each pond and raise the organic load by a factor of 10. The only problem here lies in the effluent quality. If the organic load per unit volume of the oxidation pond is raised, the microbial population will also be raised. If a low B.O.D. effluent is to be produced, these microorganisms will have to be removed from the effluent. This can be accomplished by dropping the organic load in the last pond in the series to current levels and using it as a polisher after speeding the process up in the earlier steps.

A single circular pond at a very low organic loading is by far the simplest type of biological treatment other than direct discharge into the receiving stream. It has the least operation and maintenance and is suitable for use in small towns and industries where land areas are readily available at low cost. Where maintenance can be carried out at a higher level and where land costs are high, the series ponds with high-volume recirculation should be considered. Moving up the scale the load can be increased if additional oxygen can be added to the system. Once again the ingenuity of the engineer will lead to the best possible solution of how to add oxygen to the wastes. A stream accomplishes maximum oxygen transfer by rapidly flowing over the surfaces of stones in a thin film. Recirculation of the pond contents over a stone surface in a thin film with a discharge into the pond can be used as a means of transferring oxygen at low cost if the terrain is suitable. Mechanical aeration devices at the recirculation pump or in the pond itself could also be used. Regardless of how the oxygen is transferred, the fundamental biochemistry remains the same.

Oxidation ponds have received considerable attention recently in the northern areas where the surface of the pond is frozen solid for several months at a time. If the water surface is covered and light penetration is retarded by snow cover on the ice, the oxidation pond shifts from an aerobic device to an anaerobic device. With very low organic loadings the effluent deterioration is relatively slow due to the long retention period. When the ice melts and the surface is once again agitated, there will be a short period when odors will be released. Aside from the reduced efficiency in the winter months and the short period of nuisance odors, oxidation ponds have given good operation in the Northern Plains states.

Future of Oxidation Ponds

Oxidation ponds are just beginning to be used in this country as a means of treating wastes. There is no doubt that in the Southwest and the Middle West where land is available the oxidation pond will see greater and greater use. It has the limitation of requiring large land areas for complete stabilization but the advantage of low operating maintenance. The key to the oxidation pond lies in the removal of the microorganisms which are produced as a result of metabolism of the wastes being treated. A rapid and complete microbial removal method will result in reducing the size of the oxidation ponds and is an item for future research.

Recently, oxidation ponds have been used as a means for the disposal of digested sludge and even excess sludge from activated sludge systems. In both cases, the oxygen demand of the material being disposed of is relatively low per unit volume of the oxidation pond and good results have been obtained. The major fraction of the digested sludge is inert solids and at periodic intervals it will be necessary to remove solids if the pond is not to fill up with solids.

Industrial waste disposal by oxidation ponds has begun to receive more attention, especially from small industries which cannot afford to have a skilled operator or complex treatment devices. Recently, in Rhode Island a small milk bottling plant converted an adjacent shallow pond into an oxidation pond for the disposal of wastes and eliminated construction of an expensive activated sludge system. Other industries will undoubtedly evaluate oxidation ponds when considering waste treatment systems and it is important that the engineers know how the oxidation pond works and what it will really do.

CHAPTER 23

Anaerobic Digestion

The satisfactory disposal of the concentrated organic solids removed from sewage in the primary sedimentation tanks and the excess biological solids from trickling filters or activated sludge is often brought about by an anaerobic biological treatment process commonly referred to as anaerobic digestion. The basic problem with these concentrated organic solids is the simple fact that they cannot be readily dewatered.

In the absence of air the anaerobic bacteria break down the water-binding organic solids and produce a reduced sludge mass which can be readily dewatered to a stable solid. In spite of recent technological advances, the anaerobic digestion process is still the most complex and sensitive of all the biological waste processes. Most sewage treatment problems stem from anaerobic digestion. In view of the cost of the anaerobic digester as compared to the other treatment units, one cannot help but wonder why more research has not been done to improve the process and reduce the economics of anaerobic digestion to a more reasonable level.

Description of Anaerobic Sewage Sludge Digestion

The basic treatment unit of the anaerobic digestion process is merely a tank, usually a circular concrete tank with or without a cover. In those tanks without a cover the sludge mass at the top of the tank will dry out quite quickly, forming a rather solid cover. The city of San Antonio had open digesters back in 1953 with crusts 7 to 10 ft thick. Current practice is for either fixed or floating covers on the digester to prevent such a thick dry crust as that formed at San Antonio. Figure 23-1 shows a fixed cover digester with earth insulation.

Initially the tanks were designed to hold the sludge solids for several months while the microorganisms slowly brought about digestion. As the volume of solids increased, there was a demand for a more rapid process. The addition of heat resulted in increased biological activity with a shortened digestion period. Mechanical mixing and gas mixing followed

247

FIG. 23-1. Anaerobic digester with fixed cover and insulated to full depth with soil at Waxahachie, Tex.

as means of speeding up the process. It was soon possible to accomplish in 30 days what had taken 4 to 6 months.

The modern-day anaerobic digester is still a throwback to antiquity as far as science is concerned, but it is the advanced design as far as the sanitary engineer is concerned. Figure 23-2 shows a schematic diagram of the modern digester. The raw solids are pumped through an external heat exchanger to raise the temperature prior to introduction into the digester.

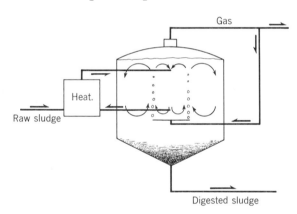

FIG. 23-2. Schematic diagram of an externally heated anaerobic digester with gas mixing.

The loss of heat from the digester requires that the contents of the digester be slowly recycled through the heat exchanger to keep the digestion tank contents close to $35°C$. Originally, it was thought that the use of the external heat exchanger with its constant sludge recycle would keep the digester contents completely mixed in the upper layer. This has since been shown to be untrue, as a pattern of shortcircuiting develops.

To keep the digester contents in motion, there is the choice between mechanical mixing and gas mixing. Properly designed mixers of either type will produce the same results. The major problem with sludge mixers is the lack of well-designed equipment. It has been indicated that gas mixing has a catalytic effect on methane production but there is no scientific basis for this. In single-stage digesters, i.e., only one digester, the mixing is usually confined to the upper volume of the digester. The lower volume is allowed to remain quiescent so that the denser digester sludge can separate from the lighter undigested sludge. In two-stage digesters the mixing is usually complete in the first-stage digester with quiescent conditions in the second-stage digester. Once the microorganisms have finished digesting the organic solids, the sludge is removed from the digester and dewatered by gravity or vacuum filtration. The dried sludge is then disposed of on the land where microorganisms break down the small quantity of organic material left.

Fundamental Theory

The complex organic solids are unavailable to the microorganisms as long as the solids remain insoluble. The initial attack is brought about by extracellular enzymes elaborated by the bacteria. These enzymes hydrolyze the complex solids to simple soluble compounds which the bacteria can utilize. The cellulose and starches are hydrolyzed to the simple sugars, while the proteins are broken to the amino acids. Only the fatty acids are not attacked by the extracellular enzymes.

The bacteria begin to metabolize the organic matter in the same pattern as indicated in previous chapters. In the anaerobic environment the bacteria do not have an unlimited hydrogen acceptor and the extent of metabolism is definitely limited. The bacteria must utilize a portion of the organic matter being degraded as their hydrogen acceptor. This results in the production of equal molar quantities of reduced and oxidized organics. The balance between oxidation and reduction is basically a function of the chemical oxygen in the organic matter being decomposed. The carbohydrates with a 1:1 ratio of C:O are the most completely decomposed materials, while the long-chain fatty acids are not touched initially. The pattern of metabolism leads to the formation of acids, which results in depression of the pH as the concentration becomes sufficiently high. The build-up of acid end products has resulted in the first phase of the digestion process to be known as the acid phase.

The high acid concentration retards further bacterial metabolism because of the low pH and because of the build-up of an end product in the system. Essentially the biological system approaches equilibrium. A second group of bacteria develops which can utilize the organic acids. As the second group of bacteria increases, the acids are metabolized to carbon dioxide and methane. Metabolism of the amino acids results in liberation of ammonia, which in turn neutralizes a portion of the remaining acids. In this way the pH rises to a more favorable level for bacterial growth. This second phase of digestion is known as the methane phase.

The methane fermentation quickly lowers the excess acids and permits further degradation of the more complex organics. The fatty acids remain essentially untouched until a healthy methane bacteria population has been produced. The methane bacteria turn their attention to the fatty acids and break them down. While the acid bacteria could not touch the fatty acids for lack of a suitable hydrogen acceptor, the methane bacteria make use of carbon dioxide as their hydrogen acceptor. The methane bacteria break the fatty acids down into simpler acids by beta oxidation with carbon dioxide as the hydrogen acceptor and water as the oxygen donor. The acetic acid which has been formed is broken down to methane and carbon dioxide by direct metabolism. The metabolic process is shown schematically in Fig. 23-3.

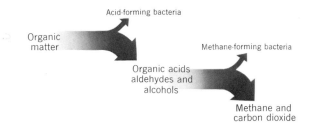

FIG. 23-3. Schematic representation of acid phase and methane phase of anaerobic digestion.

As long as the digester is able to maintain a balanced bacterial population of acid formers and methane formers, there are few operating problems. The organic matter added to the digester will be quickly converted to methane and carbon dioxide. Unfortunately, a sudden addition of a large quantity of readily degradable organic matter quickly results in the formation of excess acids with its resultant depression of pH and bacterial activity. Because of the sensitivity of the methane bacteria to slight decreases in pH, the determination of volatile acids is considered one of the most important tests in digestion operation. Actually, the volatile acids are an indirect measure of trouble. In sewage digestion a rise in volatile acids to above 2,000 mg/liter will normally depress the pH to the point where trouble is just around the corner. The relationship between volatile acids and digester problems has caused some people to place the blame directly on the volatile acids. Laboratory studies have shown that the volatile acids in themselves are not toxic to the methane bacteria. Under the proper conditions it is possible to operate a digester with good gas production at 20,000 mg/liter of volatile acids. At 20,000 mg/liter of volatile acids the methane bacteria are not limited by the acids as long as the pH is maintained about 6.5. The rate of metabolism by the methane bacteria at this concentration of volatile acids is limited by the soluble cation concentration required to neutralize the volatile acids to the desired pH level.

The battle over liming digesters to maintain the proper pH level was a hard and bloody struggle. The literature testifies to the ferocity of the contest. Unfortunately, it is impossible to purge the literature of the incorrect concept and the battle is opened anew for a brief flurry at periodic intervals. There is no question that liming is important in maintaining a suitable pH in the digester during initial startup or during periods of excess feeding. It is most fortunate that lime is the cheapest and best neutralizing agent. It has been shown that one of the limiting factors in methane fermentation is the soluble cation concentration. The monovalent cations such as sodium, potassium, and ammonium are all readily soluble, while of the divalent cations magnesium is partly soluble and calcium is the least soluble. The poorly soluble calcium removes the excess ions from solution and allows them to come back into solution as needed for neutralization.

In normal sewage digesters the digestion period has not been pushed to the point where the cation concentration is the limiting factor in the process, but there is reason to believe that as the digestion period is shortened the cations will be of extreme importance, especially ammonium ions. In the anaerobic digestion of certain industrial wastes, the cation concentration could well limit the process efficiency. A good example of this is in the effort of farmers to process manure and urine. The manure can be readily fermented by itself, but the addition of the urine increases the ammonium-ion concentration above the limiting level and retards methane production.

Types of Microorganisms

The microorganisms in an anaerobic digester are highly specialized bacteria. The fungi and the protozoa are aerobic microorganisms and cannot live in the anaerobic environment; and yet periodic isolation of fungi from sludge removed from a digester has caused some people to wonder if the fungi are not important. Actually, the fungi spores and the protozoa cysts are unaffected by the digestion process and can be isolated from digesting sludge by means of aerobic isolation techniques. The spores are merely resting as they pass through the digester on their way to a more suitable environment.

The two groups of bacteria which can live in the digester are the facultative bacteria and the obligate anaerobic bacteria. The acid formers are made up predominantly of facultative bacteria, with a few strict anaerobes. The ease in growth of the facultative bacteria give them an edge over the strict anaerobes much in the same way that they have an edge over the strict aerobe in activated sludge. It is interesting to note that the predominant bacteria in activated sludge and in one phase of anaerobic digestion are one and the same group of bacteria. Various species of *Pseudomonas, Flavobacterium, Alcaligenes, Escherichia,* and *Aerobacter* contribute to the acid production.

The methane formers are a small specialized group of bacteria that are obligate anaerobes. The methane bacteria have been so difficult to isolate and to study that very, very little is known about the individual microorganisms. Those bacteria which have been isolated belong to the genera *Methanobacterium, Methanosarcina,* and *Methanococcus.*

There is a third group of bacteria which occur in anaerobic digester, the *Desulfovibrio.* The importance of the *Desulfovibrio* depends upon the sulfate concentration of the digester. It suffices to review the fact that they are strict anaerobes which utilize sulfates at their hydrogen acceptor with the production of hydrogen sulfide as the reduced end product. In domestic sewage the sulfate reducers are not a significant part of the bacterial population, but in industrial wastes such as those high in calcium sulfate, anaerobic digestion would pose a definite problem.

Theory of Operation

In view of the fact of the biological simplicity of anaerobic digestion as compared to the aerobic processes, one can only wonder at why there are so many problems with the anaerobic digestion systems. It is the author's opinion that the problem lies with the population dynamics of the system. Current practice results in a starved biological system which is not able to handle sudden shifts in organic load because of the low bacterial population.

The basic metabolic concepts have considerable importance in understanding the results which can be expected from anaerobic bacteria. The fact that the anaerobic bacteria must metabolize more organic matter than the aerobic bacteria means that if we maintain a constant bacterial population the anaerobic system will have to be fed five times as much organic matter as the aerobic system. In current digester design practice a sewage sludge digester is designed on the basis of 0.1 lb volatile solids per cubic foot per day. Since 40 to 50 per cent of the volatile solids in sewage sludges are not biologically decomposable, the biological loading is actually only 0.05 lb of volatile solids per cubic foot per day. On a comparative basis with activated sludge the equivalent 5-day B.O.D. loading on the anaerobic digester is only 50 lb/1,000 cu ft against 40 lb/1,000 cu ft for conventional activated sludge. Considering the fact that anaerobic digesters are operated with a long retention period at high solids levels, it is not surprising to find that the mass of active micro-organisms is extremely small, generally less than 1 per cent of the total solids removed from the digester. With such a small mass of active microorganisms it is not surprising to find that there is little biologically oxidizable organic matter discharged from the digester. The major source of B.O.D. in digester supernatant is due to nitrification, i.e., the oxidation of the excess ammonia nitrogen released during the decomposition of the proteins. The low microbial population makes the anaerobic digester very susceptible to upsets from slight shock loads.

Considering the microbiology of anaerobic digesters and the biochemical reactions they bring about, it should be possible to postulate how a digester should be constructed to yield the maximum stabilization for the least effort. The few changes in anaerobic digesters in the past few years have shown that by making the environment more favorable it has been possible to reduce the digestion period down to as low as six days. Further utilization of basic principles should result in even shorter detention periods.

Population Dynamics. The key to any biological waste treatment problem lies in the establishment of an adequate biological population. The bacterial population must be at a maximum if the digester is to work at its most rapid rate and if it is to absorb shock loadings. A low bacterial population is unable to do a satisfactory job because the response lag of the bacteria to the food becomes too long. One of the major fallacies in biological engineering problems is the calculation of unit sizing and then doubling the size as a safety factor. Excess capacity does not act as a safety factor, but rather acts as a retarding factor since it reduces the microbial population below that required for good stabilization. Waste treatment systems should be constructed in modular units and operated as close to maximum capacity as possible. When the load increases in excess of the units provided, a new modular unit can be started and raised to full load. In this way treatment plants can be constructed and operated on a shorter term basis than currently being practiced. The economics of such an operation is immediately obvious.

The anaerobic digester should be operated as a complete mixing type of treatment unit in order that the microbial population can be kept relatively uniform both in total mass and in species. This means uniform feeding and complete mixing at all times in all points of the tank.

Uniform Feeding. One of the prerequisites of good digestion is the uniform feeding of the organic matter so that the microorganisms are kept in a relatively constant organic concentration at all times. Constant feeding helps to eliminate shock loadings and hence sudden increases in volatile acids. Most sewage plants are not set up to handle uniform feeding of sludge to the digester. Since uniform feeding of wastes will result in uniform displacement of solids from the digester, there will have to be some rather drastic changes in the concepts of digester design to provide for uniform feeding. Instead of a single tank unit there will have to be a two tank unit, one for digestion and one for sludge separation. The units would resemble an activated sludge system rather closely, with solids being returned from the sludge separation unit to the digestion unit. The sludge separation unit may well be a flotation unit rather than a gravity separator or even a combination of both, depending on the sludge characteristics.

Complete Mixing. Once the organic matter has been added to the digester, it must be completely dispersed to all points in the tank. This can be accomplished only with high-speed mixing. Mixing is partially art and partially science. In sanitary engineering, mixing is art at its crudest stage. The mechanical mixers and the gas mixers that are currently available on the commercial market are not capable of furnishing the type of mixing required for rapid digestion. The digester contents must be mixed to the same extent that mixed liquor is in activated sludge.

At present the author favors gas mixing in the same manner as activated sludge with continuous-gas addition at the bottom of the tank either over the entire tank bottom or along the tank periphery. The reason for this is due to the simplicity of the equipment required and the ease of maintenance. With current digestion tank design mechanical mixing will not move the sludge completely without several units and increased maintenance problems as a result of multiple units.

Current digestion tank mixing is done intermittently, since a portion of the digester is used for sludge separation. It is not necessary to utilize the reaction vessel as a sludge separating unit and it is more expensive to do so. Separate sludge separation should become a common unit of anaerobic digestion systems as soon as uniform feeding and complete mixing become more widespread.

Temperature Control. The rate of biological reactions is clearly a function of temperature. This has been amply demonstrated in anaerobic digestion where heat is added to elevate the tank temperature to between 33 and 37°C. One of the most important advances in heating digesters has been the development of the external heat exchanger to replace the internal hot-water coils. With internal hot-water coils the heat tended to dry a layer of sludge around the pipe coils so that a layer of insulation built up and slowly reduced heat transfer. The external heat exchanger draws the sludge from the digester, raises its temperature, and then puts it back in the digester. Being external it is easier to maintain and with rapid liquid movement through the pipes there is no build-up of insulating layers of sludge on the pipe walls. It was originally thought that the external heat exchanger would provide the degree of mixing necessary for rapid digestion, but the mixing is not complete enough.

One of the things that external heat exchangers opened up for reconsideration was the case of thermophilic digestion. Early research had shown that thermophilic digestion at 55°C was much more rapid than at 35°C, but it was not practical because of the high rate of sludge drying on the heating pipes. With external heat exchangers and high sludge recycle rates it is possible to raise the temperature of the digester to 55°C, but no one seems to be interested in trying thermophilic digestion. Once again the engineer is content with his current equipment and processes and sees little reason for changing them even to a more efficient system.

Limiting Factors. The hopes of anaerobic digestion sound so good that it appears from a theoretical viewpoint that the sky is the limit. Needless to say, some point a little closer to earth is more likely to be the limit. It appears at the present time that one of the major limitations in high-rate sludge digestion will be the soluble cation concentration. As the organic load is pushed higher and higher, there will be a build-up in soluble cation concentration unless more dilute sludges are treated. In domestic sewage the prime cation will be an ammonium ion. Unless the ammonium ions can be kept at a low concentration by being blown from the digester or removed, there is every reason to believe that ammonium ions will set the upper limit of digestion. What the upper limit of ammonium-ion toxicity will be cannot be said at this time since no digester has been pushed to the ammonium ion limit. Only time and further research will give an answer to this problem.

Another problem lies in the solubility of the organic compounds being biologically degraded. In order for a compound to be metabolized, the bacteria must break it down to soluble compounds that will diffuse through the cell wall. The proteins do not pose a problem from solubility, but cellulose and fats do pose a problem. Cellulose is very difficult to degrade biologically and only highly specialized bacteria can do it. As long as there is a very high population of active cellulose fermenters in the digester and as long as there is complete mixing to bring the maximum concentration of cellulose in contact with the microorganisms, cellulose should not prove too difficult. Fats pose a different problem. Fats tend to react with each other to form large balls of insoluble materials. The large fatty balls give a minimum of surface for bacterial action. Since the bacteria must be in direct contact with the fat, it is essential that the fat particles be dispersed as completely as possible to form minute particles with the maximum surface area. It may well mean the development of a high-speed dispersing nozzle in the sludge recycle system or it may well be that a unitized fat-separation system will have to be developed to remove the fat and dispose of it by another method, such as incineration. Once again here is a problem for future research.

The last problem limiting the extent of high-rate digestion is what is not known. There is so much to be learned about these microorganisms and their biochemistry that it is not possible to state all the limiting factors in their metabolism. Thus future research may well uncover a new limiting factor which will put a lid on anaerobic digestion at some point below our present expectations.

Cesspools and Septic Tanks

The oldest form of anaerobic digestion is the cesspool where all the sewage is placed into a covered pit dug into the ground and allowed to undergo anaerobic metabolism. The anaerobic bacteria convert some of the solid organic matter into liquids which diffuse out into the soil for further stabilization. The bacteria in the cesspools are the same as in the sewage sludge digester, namely, acid formers and methane formers. The lack of controlled operations prevents the methane bacteria from being established quickly and they do not play as important a role as in the sewage sludge digestion system. The major action of the cesspool is hydrolysis by the acid formers so that the organic matter is carried from the cesspool into the soil where the aerobic bacteria finish metabolizing the soluble organics.

It is essential that the cesspool have porous walls since diffusion of the liquid is the key to good operation. Normally the cesspool walls are lined with large field stones with open joints. The accumulation of solids is such that the cesspool must be cleaned at regular intervals. Chemical solvents are often required to clean the grease from the stones and unclog the leaching surface. The cesspool is a very poorly operating treatment system and must be used with extreme care. The current tendency is displacement of the cesspool with the more efficient septic tank.

The septic tank is designed to produce the same results as the cesspool except that a larger leaching field is used to disperse the liquid effluent into the soil. The septic tank is normally constructed of concrete or steel with an average retention period of two to four days' sewage flow. The tank is baffled so that the heavy solids which settle out and the scum which floats on the liquid surface do not discharge into the effluent (Fig. 23-4). Like the cesspool the septic tank fills up with settleable solids and scum and must be cleaned out. Regular septic tank maintenance requires cleaning every three to five years. Failure to clean the septic tank regularly will result in the discharge of solids to the leaching field and clogging of the field.

Biologically, the septic tank operates the same as the cesspool. The slow build-up of methane bacteria prevents much stabilization of the organic matter so that hydrolysis is the main action. The fact that the settled solids and the scum are removed from the main sphere of action prevents much change in these materials. Most of the microbial activity occurs in the intermediate liquid zone where the products of metabolism as well as the microorganisms are periodically washed out into the leaching field.

The leaching field is an integral part of the septic tank system which must be carefully designed. The size of the leaching field is dependent upon the soil characteristics. In a porous sandy soil, the leaching field can be small while in a wet, clay soil with little absorption capacity, the leaching field will be large. The major problem with septic tanks lies in those soils which have poor drainage characteristics. The effluent will not receive the proper aerobic treatment and will rise to the surface only partially treated, causing a definite health hazard. Although some sanitary engineers feel that septic tanks should not be allowed to be built, there

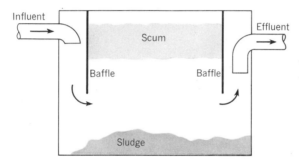

FIG. 23-4. Schematic cross section of a septic tank.

are many areas where septic tanks operate very satisfactorily and are the best solution to the sewage disposal problem. The problem is not with the use of septic tanks, but the misuse of septic tanks by engineers. A septic tank will function properly only if designed and operated properly.

Imhoff Tank

The Imhoff tank was designed to treat the solids from domestic sewage anaerobically in the same tank used for sedimentation. A cross-sectional sketch of the Imhoff tank is shown in Fig. 23-5. The sewage solids settled from the liquid into the lower compartment where they underwent anaerobic metabolism from both the acid formers and the methane formers. The gas produced by the methane bacteria was discharged through gas vents to the air. Needless to say, scum also collected in the gas vents. Figure 23-6 shows an Imhoff tank. The biological development is much better in the Imhoff tank than in the cesspool or septic tank for the rather simple reason that the food for the methane bacteria and the methane bacteria themselves are not being continuously removed but rather are being allowed to slowly build up. The end products of the acid formers remain in the sludge compartment sufficiently long for the methane formers to grow up and produce a considerable quantity of gas.

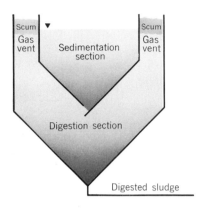

FIG. 23-5. Schematic cross section of an Imhoff tank.

The Imhoff tank is usually reversible as far as flow is concerned. The solids usually settle out at the head end of the sedimentation tank and soon fill the sludge compartment at the head of the tank. The reversal of flow through the tank permits the sludge to be distributed more uniformly throughout the entire tank. The normal design for the sludge compartment is for six to twelve months. The sludge which is removed from the sludge compartment is usually well digested and can be dewatered on drying beds. The major drawback with the Imhoff tank is the cost of construction and the slowness of digestion. Separate, heated digestion has all but eliminated the Imhoff tank in sewage treatment for large installations. There are some small plants which can economically use Imhoff tanks.

Future Developments

An editorial in one of the leading British technical journals in the field of sanitary engineering took Dr. C. N. Sawyer to task for daring to push the frontier back on anaerobic digestion. The editorial felt that current design concepts were adequate and as long as they performed satisfactorily there was little need for shortening the digestion period since that would only create further problems for the sanitary engineer to solve. This editorial was not expressing an attitude common to England, but rather was expressing the feeling of the majority of the engineering profession. Most engineers are satisfied with current design concepts and do not want to change. There is nothing better than *status quo*. But fortunately there are a few people who are real engineers endowed with the true engineering spirit of doing the best possible job for the least cost. These few engineers will be the ones to show how to improve anaerobic digestion and will bring about improvement in spite of the field.

The future developments in anaerobic digestion currently point to complete waste treatment by anaerobic systems. The higher loading rates favor anaerobic treatment over aerobic treatment, but current effluent quality favors the reversal. Part of the problem with anaerobic treatment lies in separating the dispersed cells from the final effluent. In aerobic systems the protozoa assist with clarification but there are no assisters in anaerobic systems. Actually, the anaerobic system can produce an effluent of as good quality as the aerobic system once the dispersed cells are removed. Here again is the problem for research.

With sewage sludge and the large concentration of inert solids it contains there is reason to believe that digestion periods will be made very

FIG. 23-6. Imhoff tank at Marlboro, Mass., showing the gas vents.

short with a high degree of solids wasting so that the active mass will increase considerably above 1 per cent. There is no doubt that the increase in active fraction will require a change in dewatering practices, but it is doubtful if the changes will be major.

Anaerobic digestion is the uncharted wilderness in sanitary engineering. It holds the promise of vast expansion and new waste treatment processes and at the same time will no doubt offer pitfalls and problems that make aerobic treatment problems seem easy by comparison. The future is so full that it is impossible to state where the end lies and what exactly the full potentialities are.

CHAPTER 24

Refuse Disposal

The disposal of refuse by microorganisms is one of the oldest biological waste treatment systems. Man has buried unwanted refuse for centuries and the microorganisms have broken the complex organic compounds into simple compounds which have been reused by higher plants and converted back into complex organic matter. Today there are two classes of refuse disposal by microorganisms: sanitary landfill and composting. Sanitary landfill is the most widely used of the two processes, but composting is obtaining greater recognition and will play an important role in conservation of natural resources for future generations.

SANITARY LANDFILL

The sanitary landfill can be described as engineered burial of refuse. A trench is excavated to a depth of 10 to 15 ft deep and approximately 20 ft wide. The refuse is dumped from trucks into the trench to a depth of 6 to 10 ft, where it is compacted by a bulldozer and covered with 4 to 6 ft of dirt. The microorganisms slowly decompose the organic matter to stable compounds. The microbial degradation is at a very slow rate when compared with the other biological treatment systems.

Most of the breakdown of the complex organic matter is brought about by facultative bacteria and fungi. The facultative bacteria work under anaerobic conditions to hydrolyze the complex organics to the simpler water-soluble organic acids. These organic acids diffuse through the soil where fungi and other bacteria aerobically metabolize the organic matter to carbon dioxide and water. Often the methane bacteria build up with the discharge of methane through the soil. As the methane diffuses through the soil, aerobic methane bacteria develop which can utilize a portion of the methane although much of the methane will be lost to the atmosphere. When too much organic matter is buried without consideration for methane production, there is a potential fire hazard.

Moisture is essential for biological degradation. If the buried refuse contains less than 60 per cent moisture, it will be difficult for the bacteria to grow and degrade the organic matter. Often in low-moisture-containing refuse the fungi start metabolism aerobically with the production of carbon dioxide and water. The production of water as an end product of metabolism permits the bacteria to grow and more water is produced as a result of their metabolism. Ground water also assists in keeping the refuse moist enough for microbial activity and for diffusion of end products to more aerobic zones. On the negative side ground water fills the air voids and prevents aerobic metabolism.

COMPOSTING

Composting is very similar in its action to a sanitary landfill but is a controlled microbial reaction yielding a stable end product much sooner. The purpose of composting is to remove the readily degradable organic matter from the refuse and to produce a stable material that can be used to recover waste land or to grow food crops. In areas of the world where the population per unit of land mass is very high, such as in Japan, China, and India, composting of refuse is practiced to recover the nutrients as fertilizer. In the countries rich in mineral fertilizers, composting has never been practiced to any extent. In this country composting of refuse has never been successfully used in a large scale for a long period of time, although there have been limited operations for several years. As the need for waste-land recovery increases, there will be greater emphasis on composting.

Types of Composting

There have been two types of composting used thus far, the open windrow and the mechanical.

Open-windrow System. This method of composting consists in placing the refuse in piles approximately 4 ft high and 8 ft wide. The moisture content of the compost is adjusted to approximately 60 per cent and biological activity is allowed to begin. After several days the waste heat from the microbial reactions will build up in the compost pile. The microbial populations shift from mesophilic to thermophilic reactions, as the temperature will rise up to 70°C. The pile is turned before the temperature rises too high and the moisture content adjusted to 60 per cent. Turning of the pile permits cooling and assists in aerating the pile so that the metabolism remains aerobic. The temperature of the pile will once again rise to 70°C and the pile must be turned again. After several turns the temperature will fail to rise back to 70°C and will begin to fall off. The failure of the compost to increase in temperature has been used as the means of determining that the compost has been stabilized. The open-windrow composting process takes approximately six to ten weeks to be complete.

Mechanical System. In an effort to speed up the composting process mechanical devices have been constructed to turn the compost continuously, adjust the moisture content, and add air. Some of the mechanical units have been constructed vertically, while others have been built horizontally. Although the mechanical devices are all proprietary devices, they operate on the same principles as the windrow composting. The only difference is the fact that the time required to produce the stable compost is only three to six days.

Microorganisms in Composting

Composting is carried out in a semimoist condition which favors the fungi and actinomycetes as the predominant group of microorganisms, with bacteria assisting in the upper moisture ranges. Very little work has been done on the microbiology of composting other than to show that natural mixtures of bacteria actinomycetes and fungi will produce a compost just as easily and just as fast as the so-called special pure cultures which some people feel are necessary.

The fungi which predominate in the compost will be dependent upon the organic matter being decomposed. Various species of *Penicillium, Rhizopus, Aspergillus,* and *Mucor* will be found in the compost. The bacteria found in the various waste treatment processes will also be found in the compost with thermophilic species predominating at the higher temperatures. The mesophilic bacteria will usually be of value only in starting the compost.

Temperature

Temperature is one of the chief control methods for composting. A thermometer is placed in the interior of the compost mass so the temperature can be determined at regular intervals. The temperature of optimum thermophilic composting is around 65°C. If there is too much moisture in the compost, there will not be enough heat to raise the temperature to optimum. Thus it is important that the moisture is controlled closely. The primary microbial reaction is an aerobic thermophilic reaction so that sufficient air must be present to keep the system aerobic. Too much air passing through the composting mass will remove too much heat and the temperature will not rise to optimum. Not enough air and the reaction will become anaerobic and odors will be produced. As we have already indicated, when the temperature of the compost pile does not rise, the compost can be considered as complete.

Moisture

There must be sufficient moisture to permit the microorganisms to hydrolyze the complex organic compounds into simpler compounds which diffuse through the cell wall. Too much moisture results in lowered temperatures and anaerobic conditions, while too little moisture prevents metabolism. The range of moisture appears to be from 40 to 70 per cent. If the refuse being composted is too dry, water is added and if it is too wet, dry material such as paper is added to absorb the excess moisture. The optimum moisture content has been found to be 60 per cent.

Hydrogen-ion Concentration

The pH of the composting mass must also be carefully controlled. The normal metabolic reactions produce organic acids which will lower the pH unless sufficient buffer is present to hold the pH at the desired level. Lime is generally added for pH control because of its price, ease of handling, and low solubility. Being poorly soluble, it is taken up by the acids as needed. Since much of the metabolic activity is brought about by fungi, the compost can operate at lower pH levels than the other biological treatment systems. Optimum pH is at 6.5, but satisfactory results can be obtained from pH 4.5 to pH 9.5.

Nutrients

The formation of microbial protoplasm which is the net result of the degradation of the organic matter in the refuse requires a definite quantity of nitrogen and phosphorus as well as trace elements. The compost generally supplies all the necessary trace elements but often there is a deficiency of nitrogen and phosphorus. The nutrient requirements have generally been expressed in terms of a carbon:nitrogen ratio. The optimum C:N ratio is 20:1, while the C:P ratio is about 100:1. A nitrogen deficiency requires that an ammonium salt be added to the compost.

Air

In mechanical composting air is added to the composter continuously. It has been found that 10 to 30 cu ft of air per day per pound of volatile solids is required for the metabolic reactions. There must be some excess air supplied in order to ensure aerobic conditions. Air in excess of 30 cu ft per day per pound of volatile solids has been found to result in drying of the compost mass and loss of excess heat.

Particle Size

The size of the organic particles is very important in the rate of the composting reaction. It is essential to have the maximum surface for the microorganisms to come into contact with. This means the particle size should be a minimum. Yet, if the particles are too small, there will not be sufficient voids for maintaining the system aerobic. It has been found that ball-mill grinding to a particle size of ½ in. will produce satisfactory results in mechanical composting, while particle sizes in windrow composting should be between 1 and 1½ in.

Materials Composted

All types of solid organic matter can be composted under proper conditions. Waste sawdust from sawmills can be composted but the highly specialized microorganisms involved in cellulose metabolism require close control from both a pH and a nutrient standpoint. Cotton hulls from ginning operations have been successfully composted as have corn cobs and other waste agricultural products. The major emphasis in the past decade has been on composting municipal refuse.

Municipal refuse represents the solid waste products from the community in question and will vary from place to place. Analyses of municipal refuse at Berkeley, Calif., by the University of California indicated that the refuse contained 9.8 per cent by weight tin cans, 11.7 per cent bottles and glass, 1.6 per cent rags, 0.9 per cent metal, 7.6 per cent inorganics, and 68.4 per cent compostable material. One of the major problems with composting is separation of the noncompostable material from the compostable material. Recent equipment developments have greatly improved the problem of separation of compostable and noncompostable materials. The refuse can be coarse ground and heavy metallic objects removed in a single operation. Glass is pulverized to a fine powder and retained with the compostable organics. Light ferrous metallic objects, such as tin cans, are separated by means of electromagnets. Two-stage grinding usually produces a more uniform material for composting than single-stage grinding.

One of the major problems in handling municipal refuse is the fact that microbial activity is well advanced before the material reaches the composting area. The anaerobic activity is usually sufficient to cause an odor problem in the immediate vicinity of the composting area. It is essential that the time for preparation of the compost be held to a minimum so that the odor nuisances will also be held to a minimum.

Compost Analyses

The finished compost is a dark, stable humus material with little fertilizer value but some value as a soil conditioner. It is the lack of fertilizer elements that has handicapped the use of composting as a means of refuse disposal. Yet the reclamation of arid areas or clay soils by the use of composted materials has potentialities as the demand for land reclamation increases. Examination of an analysis of compost from municipal refuse shows 1.2 per cent nitrogen, 1.1 per cent phosphorus pentoxide, 0.5 per cent potassium oxide, 24.4 per cent carbon, and 39.4 per cent ash. The fact that most of the nitrogen and phosphorus is tied up in microbial protoplasm means that these elements will be released slowly into solution where plants may take them up for their growth. The real advantage for compost lies in its use as a base for chemical fertilizers, thereby producing a combined organic-inorganic fertilizer with good soil-conditioning properties.

Future of Composting

In the United States there will be little increase in composting during the next 50 years simply because the economics of composting is not favorable. Materials handling is too much of a problem with the high cost of labor. It is possible to produce more compost in a single large city than can be consumed in an area several hundreds of miles from that city. At the present time chemical technology has produced better fertilizers and soil conditioners at lower prices than compost can be produced and sold. Until the population of the United States reaches the level where the demand for food, as currently produced on existing farms, is greater than the production, compost and sewage sludge will have little economic importance in this country. Eventually, there will be a demand to reclaim arid regions and compost can be used in such reclamation, but this problem is several generations away.

In some areas of Europe and Asia the lack of adequate mineral resources makes the use of compost as a fertilizer an economic necessity today. It is in these areas that need compost that most of the developments are going to be made.

CHAPTER 25

Radioactivity

This is the atomic era in which nuclear fission holds promise of great things to come. The nuclear developments have posed some problems for sanitary engineers with regard to waste disposal but have given them new tools with which to study the biological waste treatment processes more thoroughly than ever possible before.

NUCLEAR WASTE DISPOSAL

The disposal of nuclear wastes poses an unusual problem for the sanitary engineer in that these wastes are constantly changing in energy levels at varying rates for each radioactive element. As the radioactive elements decay, alpha particles, a helium nucleus (He^{++}), beta particles, an electron or a positron from the nucleus, or gamma rays, photons, are given off. These high-energy particles or rays can cause nuclear changes in any materials which they strike. This reaction is of importance to sanitary engineers when the radiation strikes biological materials. In biological materials radiation can produce minor changes or major changes depending on where the reaction occurs. If the radiation should strike the genes which are responsible for the transmission of hereditary characteristics, a change, or mutation, can occur in which the offspring does not have the same characteristics as the parents. In complex biological organisms, such as human beings, radiation can produce changes which are fatal as well as mutational. It is the sanitary engineer's job to see that the radiation in nuclear wastes does not reach sufficient concentration to cause damage to human beings or to microorganisms.

Decay

All radioactive elements give off radiations at fixed rates dependent upon their atomic structure. In some elements the energy is dissipated very rapidly, while in others it is dissipated very slowly. The rate at which the energy is given off is measured in terms of half life, i.e., the time required for one-half of the available energy to be dissipated (Fig. 25-1).

The radionuclide, N^{16}, has a half life of only 7.4 sec, while N^{13} has a half life of 10 min. Other elements have half lives measured in hours, in days, and in years. From a practical standpoint the radionuclides with half lives measured in less than days do not pose a problem in radioactive waste disposal, as they decay to stable elements before the wastes are

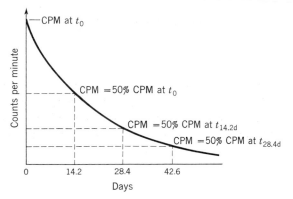

FIG. 25-1. Decay curve for P^{32} illustrating principle of half life.

discharged. The long half-life elements are the source of problems since there is no known method at this time to speed up the rate of decay of these elements.

Waste Treatment

Since the problem with radioactive waste disposal lies primarily with the radiation and since the rate of decay is fixed, there is only one known waste treatment process for radioactive wastes, holding. The wastes must be contained sufficiently long for the radionuclides to decay to a level where it is safe to discharge them into man's environment for ultimate disposal by dilution. While waste treatment is simple in theory, it is far from simple in practice. There is a limit as to how much wastes can be stored for a given period of time. Thus it is that the capacity for waste storage sets the limits for any nuclear operations.

It is impossible to impound every bit of radioactive wastes indefinitely and it is unnecessary. With sufficient dilution it is possible to allow some of the radionuclides to be discharged into streams, the ocean, or the atmosphere where they can continue their rate of decay without damage to man.

Tolerance Levels

The level of tolerance for radioactivity to which a person can be exposed to depends in part on the nature of the radiation, which in turn is related to the particular radionuclide. The Atomic Energy Commission has set certain standards of tolerance for the various radionuclides. These are reported in the *Congressional Record* and various radiation handbooks. Unfortunately, with gross mixtures of radionuclides as in most wastes it is not possible to set forth variable tolerance levels. A single general tolerance level has been set for unknown radionuclides in water at 10^{-8} μc/ml for lifetime consumption.

One of the most controversial questions today is what the tolerance levels should be. The relative newness of nuclear developments has prevented establishment of absolute tolerance levels. The net result is periodic revision of the tolerance levels. There are many geneticists who are opposed to the release of any radiation into the environment and are critical of any tolerance limits as being safe. In an effort to be safe the AEC has set the tolerance levels as low as possible even below well-established safe tolerances in order to satisfy the most critical geneticists.

Sources of Wastes

Radioactive wastes arise from the mining and processing of uranium ore, research and medical laboratories, and nuclear reactors. The mining and processing of uranium gives rise to wastes containing dilute quantities of natural radionuclides resulting from the decay of uranium. The major radionuclide in ore-processing wastes is Ra^{226} which has a half life of 1,620 years. The wastes are discharged to tailing ponds where the insoluble materials settle out and only the soluble materials are discharged to the receiving stream. At the present time the maximum limit of radionuclides allowed in the tailing pond effluents is 4 $\mu\mu$gm/liter Ra^{226} +50 per cent daughter products. Stream surveys by the USPHS have shown that some of the radioactivity is concentrated in the stream muds and in the algae growing in the stream, but that the total quantity of radioactivity discharged did not constitute a health hazard.

The rapid increase in the use of radionuclides for research and medical therapy has created a dilute waste disposal problem. The major radionuclides used as tracers include P^{32}, C^{14}, and I^{131}. In low concentrations, 10^{-7} to 10^{-8} μc/ml, these radionuclides can be disposed of directly to the sewer, provided sufficient carrier is added to prevent a large uptake by microorganisms in the sewer. With solids and concentrated wastes, disposal by burial at sea has proved a satisfactory method of disposal. Although the use of radionuclides in research and medicine will increase, the concentration of wastes will not create a major disposal problem since the concentration of radionuclides at any one point will be relatively low.

Nuclear bomb tests have produced one of the major waste disposal problems. Release of radioactive debris into the atmosphere following a nuclear bomb test has raised the level of radioactivity in almost every area of the world as a result of fallout. It has been possible to determine nuclear bomb tests by sudden increases in fallout of nuclear debris. Much of the radioactive fallout occurs with precipitation such as rain or snow. Since this rain and snow forms much of the surface water used for human consumption, the sanitary engineer has been very interested in fallout. Sr^{90} is one of the major radionuclides in fallout and is of concern because plants tend to concentrate it and pass it on to humans where it is concentrated in bone tissue. It is feared by some scientists that the increase of Sr^{90} in bone tissue will result in an increase in leukemia due to radiation damage. The effect of fallout on the genetic characteristics is currently one of the most controversial scientific and political topics.

Nuclear reactors produce two major liquid wastes for disposal, cooling-water wastes, and fuel-reprocessing wastes. There will be certain solid wastes resulting from contamination of equipment and replacement. Cooling-water wastes are usually a minor problem since the tendency in reactor design is to shift from a single-pass cooling system to a recirculating cooling system. In the single-pass system, the water is passed around the reactor once and then discharged back to the stream after suitable storage to allow the radionuclides to decay to a safe level. Most of the radionuclides in the cooling water result from induced radiation of the natural ions in the water as well as corrosion products which dissolve into the water. The magnitude of induced radiation is small and most of the radionuclides thus formed have very short half life so that storage is suitable treatment. In recirculating systems the radionuclides which result are primarily from corrosion products and are removed by demineralization.

The fuel-reprocessing wastes are the major radioactive waste problem today. The major fuel elements used in nuclear reactors consist of uranium clad in aluminum, stainless steel, or zirconium. As the fuel is used, daughter products are formed, which eventually poison the system. Much of the uranium fuel still remains in the poisoned fuel element and reprocessing is necessary to recover the uranium for further use. The daughter products which have resulted from the decay of the uranium are called mixed fission products.

When fuel elements are removed from the reactor, they are allowed to cool in storage for 90 days to allow the short life radionuclides to dissipate their energy. The fuel elements are dissolved in concentrated chemicals. Aluminum-clad fuel elements are dissolved in sodium hydroxide–sodium nitrate solution, followed by nitric acid and nitrous acid. The fuel mixture is then added to an extraction column where aluminum nitrate is added at the upper end and tributyl phosphate is added at the lower end. The uranium is extracted by the TBP while the mixed fission products contaminating the fuel are retained in the aqueous phase by the aluminum nitrate. The uranium is further processed by succeeding chemical steps. The aqueous mixed fission products form the bulk of the wastes for disposal. While it is possible to extract certain radionuclides from the mixed fission products for recovery, they generally represent unwanted radioactivity. The radioactivity runs from 1 to 4×10^2 curies/gal with an effective life of 600 years. In addition to the radioactivity the wastes are strongly acid with a very high concentration of aluminum ions and nitrate ions. The chemical nature of these wastes make them extremely difficult to treat. Because of the extremely dangerous potentiality of these wastes, they are carefully kept in storage. Considerable heat is generated by the decay of the radioactivity so that it is necessary to keep these wastes cool. Storage in metal tanks results in eventual corrosion and the need for replacement of the storage tanks by new tanks, so that the waste disposal problem represents long-term storage. At the present time it is estimated that nuclear power reactors in the United States by 1964 could produce 90×10^6 gal of mixed fission wastes each and every day of operation. It is easy to see why waste disposal acts as the rate controller on the development of nuclear power.

One of the most recent methods proposed for waste disposal has been the conversion of the mixed fission products into ceramic glazes which could be buried in the earth without special containers. Laboratory studies have shown great potentialities for disposal of aluminum-clad and stainless-steel fuel element wastes by conversion to ceramic masses.

Radioactivity in Streams

The discharge of dilute radionuclides in sewers directly to streams is of importance to the sanitary engineer since microorganisms will tend to concentrate the radionuclides in their protoplasm. Fish will consume many microorganisms and will concentrate the radioactivity in their tissue. Consumption of the fish by humans will result in transfer of the radioactivity to the critical points within the body.

The initial removal of the radionuclides by microorganisms depends upon the relation of the chemical properties of the radionuclide and the demand by the microorganisms for this element. Microorganisms will remove the radionuclides by incorporating the elements into protoplasm. A high concentration of stable isotope will reduce the uptake of the radionuclide by the microorganisms. This is the reason why the AEC requires discharge of a high concentration of stable isotope along with the radionuclides which are discharged to the sewer.

By keeping the level of radioactivity discharged to streams well below safe levels, 10^{-7} to 10^{-8} μc/ml prior to discharge, it is not possible for the microorganisms to concentrate the radionuclides to a level where it would be detrimental to humans.

Fallout

The concentration of radionuclides in surface water from fallout has been studied rather extensively. These studies have shown that there was a rise in radioactivity in surface waters following nuclear bomb tests. At no time has the radioactivity risen above the tolerance level in the reservoirs prior to treatment even though some rains have shown radioactivity as high as 8×10^{-5} $\mu c/ml$. Normal water purification processes such as coagulation, sedimentation, and filtration will remove certain of the radionuclides, but much of the radionuclides in fallout will pass through the treatment plant. Approximately one-third to one-half of the radioactivity can be removed by complete treatment. Thus, water purification cannot be counted on as a major decontamination procedure from nuclear debris in times of an emergency.

The concentration of Sr^{90} in plants from the mixed fission products in fallout has caused some scientists to be concerned with human consumption of certain foods which might be grown in areas following a nuclear devastation. There has been enough concern over the concentration of Sr^{90} in milk that warnings have been sounded to avoid consumption of milk in case of a nuclear disaster.

Biological Waste Treatment

Biological waste treatment has not proved satisfactory for concentrating radionuclides. The reason for this lies in the fact that biological waste treatment systems are in the declining growth phase and are releasing elements while removing these same elements. The metabolism of organic matter results in the production of new protoplasm with the demand for a certain quantity of chemical elements per unit of organic matter metabolized. Once the cells have been formed, they undergo endogenous metabolism with a release of the elements which formed protoplasm. The net removal of chemical elements will depend upon the rate at which the microorganisms are grown and separated from the liquid. In an activated sludge system the rate of removal will depend upon the rate of sludge wasting. There will also be competition between the stable isotope and the radionuclide. The microorganisms have a fixed demand and will remove the two proportionately. The maximum concentration of radionuclide will be removed when the total concentration of stable isotope and radionuclide is below the microbial demand for the element.

Once the cells have concentrated the radionuclides, there is the problem of ultimate disposal. If the microorganisms are added to an anaerobic digester, the cells will be destroyed and the radionuclide released back into the liquid phase with nothing being accomplished. It is possible to filter and dry the solids with ultimate volume reduction being made by incineration. The radionuclides would generally remain in the residual ash which could be processed further by leaching and chemical concentrating. Needless to say, the processing required is much greater than the value received.

RADIOTRACERS

Radiotracers have never been extensively used in sanitary microbiology although they have been used to a considerable extent in fundamental microbiology. The prime reason for this is the lack of interest in sanitary microbiology. It is believed that as interest increases in sanitary microbiology there will be an increased usage of radiotracers in research. Much of the information pertaining to fundamental microbiology will be of direct use in sanitary microbiology and offers the starting point for any tracer work.

Fundamental Microbiology

Tracers have been largely used in fundamental microbiology to elucidate metabolic pathways. Tracer studies using C^{14} have shown that bacteria metabolize glucose by two pathways. Under certain conditions with certain bacteria the Emden-Meyerhof system is used, while for other bacteria the gluconic acid shunt is used and for still others a combination of both systems. Many mechanisms of metabolism have been shown by following tagged atoms. Tagging a compound with a radiotracer permits its easy detection and separation from other compounds in the biological mixture.

Sanitary Microbiology

In sanitary microbiology, tracers have shown that in methane fermentation of acetate, the methyl carbon gives rise to methane and the carboxyl carbon gives rise to carbon dioxide. In the metabolism of the higher fatty acids such as propionic acid and butyric acid, tracer studies have shown that these acids are oxidized by using carbon dioxide as the ultimate hydrogen acceptor with carbon dioxide being the source of the methane.

Tracers were also used to follow the degradation of alkylbenzene sulfonates, the common detergents. Radioactive sulfur showed that these detergents were not completely metabolized by activated sludge. The similarities of batch-fed activated sludge systems and complete mixing activated sludge systems were shown by the use of radiotracers. Under identical conditions it was shown that the quantity of organic matter converted to protoplasm and the quantity of protoplasm decomposed by endogenous metabolism was the same. Thus an important concept in activated sludge systems was confirmed, i.e., organic matter is stabilized only by its conversion to new protoplasm. New protoplasm is constantly replacing old protoplasm. Since the old protoplasm cannot be completely metabolized biologically, it means that even in a total oxidation system whether batch-fed or completely mixed there will be a build-up of residual inert organic solids and the quantity of inert organic solids will be the same in both systems.

Tracers in the form of C^{14} in lactose have been used to determine the presence of coliforms in a few hours rather than in days. As the coliforms metabolize the lactose, the C^{14} is given off as carbon dioxide gas and easily detected.

The use of radiotracers should assist in determination of many of the fundamental mechanisms involved in the various waste treatment systems. A better knowledge of the fundamental mechanisms of the various treatment processes will lead to better treatment plant designs and operations. Tracers can also assist in the determination of complete metabolism of unusual synthetic compounds which are appearing more and more in sewage and industrial wastes. Radiotracers are a tool which can assist the sanitary microbiologists, but care must be taken not to consider radiotracers as an end within themselves. It appears that sanitary microbiology and radiotracers will both have the opportunity to grow up together.

CHAPTER 26

Milk and Food

One of the problems facing the sanitary microbiologist is control of the microbial purity of foods distributed to the public. For the most part the sanitary microbiologist is interested in the prevention of the spread of disease rather than preservation of the food products. The magnitude of the food industry today is such that there is an entire branch of microbiology devoted to food technology. While each branch of microbiology has its own separate interests, the branches have common interests in sanitation.

The development of governmental regulations on food sanitation has greatly reduced the problem of food-spread diseases so that the sanitary microbiologist finds himself primarily interested in control rather than in development. The major sources of potential health hazards are milk, other food, and restaurants.

MILK

Milk is an essential food which can easily become a source for the spread of microbial infections. The high nutrient value in milk makes it an ideal medium for the growth of microorganisms. It is essential that the microbial content of milk be controlled not only to prevent the spread of disease but also to prevent spoilage. Milk contains approximately 3 per cent proteins, 4 per cent butter fat, and 5 per cent lactose in addition to all the inorganics required for microbial growth.

Source of Microorganisms

The microorganisms in milk come from four sources: (1) the cow, (2) the air, (3) the milk containers, and (4) the milk handlers. It is essential that only milk from healthy cows is used. But even a healthy cow can cause microbial contamination if the udder is not clean. The air in the milking area can contain a large number of microorganisms unless it is kept clean and free from dust, dirt, and manure. One of the major problems is maintenance of proper sanitation in the milking areas. The containers in which the milk is placed must be thoroughly cleaned with a bactericidal cleanser after each use. The presence of butter fat can cause a thin layer of organics to adhere to the milk containers and serve as a breeding ground for bacteria unless it is removed by proper cleaning. It goes without saying that milk handlers must be healthy when working with milk. Personnel sanitation is a continuous project to ensure that contamination from this source is held to a minimum.

Types of Microorganisms

Bacteria are the primary microorganisms which grow in milk. For the most part the bacteria which predominate in milk are grouped according to the biochemical reactions which they bring about.

1. *Acid producers*—form acid from metabolism of lactose. The acid formers include the *Streptococci, Lactobacilli,* and *Micrococci.*
2. *Gas producers*—form acid and gas from the fermentation of lactose. The gas producers are primarily coliforms, although the anaerobic sporeformers *Clostridium* can also produce gas.
3. *Proteolytic*—hydrolyze the milk proteins producing an alkaline reaction. The proteolytic bacteria are primarily the aerobic sporeformers, *Bacillus.*
4. *Alkali producers*—releases ammonia during the metabolism of protein, causing the pH to rise. The alkali formers include *Shigella* and *Pseudomonas.*
5. *Inert*—all bacteria which grow but do not produce an obvious biochemical change are lumped together as inert. Since they grow, they must utilize some organics and hence are not truly inert.

Diseases Transmitted

The diseases which are transmitted through milk are grouped according to their source. The diseases of bovine origin include the following:

1. Tuberculosis—*Mycobacterium tuberculosis* var. *bovis*
2. Mastitis—*Streptococcus agalactis*
3. Undulant fever—*Brucella abortus*
4. Foot and mouth—virus
5. Anthrax—*Bacillus anthracis*

The diseases of human origin include:

1. Typhoid fever—*Salmonella typhosa*
2. Scarlet fever—*Streptococcus scarlatinae*
3. Diphtheria—*Corynebacterium diphtheriae*
4. Cholera—*Vibrio comma*

It is readily apparent from the diseases which have been transmitted through milk that it is important to remain vigilant at all times if localized epidemics are to be prevented. The fact that few diseases are transmitted through milk in this country testifies to the importance of proper sanitation in controlling disease transmission.

Pasteurization

Pasteurization of milk has played a major role in controlling the spread of pathogenic bacteria through milk. The purpose of pasteurization is not sterilization, but rather disinfection. It was found that holding milk at 143°F for 30 min resulted in destruction of the pathogenic bacteria but not all the saprophytic bacteria. This combination of time and temperature did not produce much change in the chemical nature of the milk. Longer retention periods than 30 min imparted cooked flavors to the milk. The large tank volumes required to hold the milk for 30 min caused research to be directed to shortening the time of contact. The desired bacterial kill could be obtained in shorter periods only if the temperature was raised. The higher temperatures produced undesirable tastes in the milk in most cases. It was found that at 161°F and 15 sec retention, the undesirable tastes were not produced and pasteurization was accomplished. The high-temperature–short-time pasteurization has become the standard for handling large volumes of milk. This method requires careful control, as pasteurization will not be accomplished if the milk is not entirely heated at 161°F for 15 sec. Radiotracers have been used to determine the contact period in short-time pasteurizers.

Phosphatase Test

Control of pasteurization would be difficult if it were not for the phosphatase test. Phosphatase is an enzyme present in milk which is heat-sensitive at the normal ranges for pasteurization. Holding raw milk for 30 min at 143°F destroys 96 per cent of the activity of phosphatase. It is possible to detect a 1°F temperature differential or 5 min underheating or the addition of only 0.5 per cent raw milk by the phosphatase test. The phosphatase test can also be used with the high-temperature–short-time pasteurization method.

Phosphatase is an enzyme which hydrolyzes phosphate from organic compounds. The phosphatase test depends upon the hydrolysis of a phenylphosphate ester to form phenol and phosphate. The addition of BQC, 2,6-dibromoquinonechloroamide, causes a blue color reaction to occur with the phenol. Thus the presence of phosphatase is demonstrated by a blue reaction, while the destruction of phosphatase by proper pasteurization fails to produce a blue color.

Milk Quality

It is very important that the quality of milk be determined quickly before the milk is prepared for delivery to the consumer. Since most bacterial contamination results from poor sanitation, it is possible to use the microbial content of milk as a measure of its sanitary quality. The bacteria count in raw milk is used to determine part of the grading requirements. The USPHS has indicated that Grade A milk should have less than 200,000 bacteria per milliliter, Grade B milk should have less than 1,000,000 bacteria per milliliter while Grade C milk has no bacterial limit. The bacteria counts are determined by plate counts using tryptone glucose agar with incubation at 37°C. The time of incubation has caused other techniques to be developed for estimating the microbial content of milk.

The high level of bacteria populations in the raw milk permits the use of direct microscopic counting of the microorganisms. Direct microscopic counting is made on a dried film of milk which has been stained with methylene blue. The development of accurate pipettes with a delivery of 0.01 ml has permitted fairly accurate bacteria counts to be made in just a few minutes. The standards governing bacterial counts usually permit the use of standard plate counts or direct microscopic counts.

Reductase Test

It is not possible to examine every bit of milk routinely for bacteria counts and yet it is necessary to know something of the microbial activity in the milk. The metabolism of the bacteria in milk results in the build-up of reduced metabolic products since the lack of oxygen in the milk causes the metabolism to be anaerobic. The quantity of reduced end products is in direct proportion to the total microbial population and activity in the milk. By adding an oxidation-reduction-potential (ORP) indicator such as methylene blue, it is possible to determine the microbial activity by the disappearance of the blue color which indicates reduction of the methylene blue. If the methylene blue is not reduced in 8 hr, the milk is of excellent quality. A reduction of the methylene blue between 6 and 8 hr indicates a good milk, while a reduction between 2 and 6 hr is fair. Milk which shows methylene blue reduction in less than 2 hr is definitely a poor milk.

Refrigeration

It is essential that milk be kept refrigerated at all times from the cow to the consumer except when being pasteurized. Refrigeration keeps the rate of growth of the bacteria at a minimum and maintains the quality of milk for its maximum period. Even with the best of refrigeration the microbial activity proceeds at a definite rate, ever increasing as the number of microorganisms increase. The milk eventually becomes sour due to the acid reaction brought about by the bacteria. While fermented milk products have some value, uncontrolled fermentation usually results in spoilage.

FOOD

Uncooked food can serve as a vector for the spread of disease if proper sanitation precautions are not taken. For the most part the prime diseases spread through uncooked food include trichinosis, typhoid fever, salmonellosis, amebiasis, and food poisoning due to *Staphylococcus* or *Clostridium botulinum*.

Trichinosis

Trichinosis is caused by a parasitic worm, *Trichina spiralis*, through uncooked or poorly cooked pork. The *Trichinae* exist in the flesh of the pig and are not readily observed so that pork containing the parasite can pass undetected. With proper cooking the *Trichinae* can be completely destroyed. In the United States the infection of pigs by *Trichinae* is very widespread. Efforts have been made to prevent the spread of *Trichinae* in pigs by cooking all garbage fed to the pigs. The material-handling problem in cooking garbage fed to pigs has caused some pig farmers to ignore proper cooking so that trichinosis is still widespread among pigs. It is possible to kill the *Trichinae* by keeping the pork refrigerated at 5°F for 20 days. Trichinosis is not a major problem in spite of the large number of infected pigs because of efforts to educate people to eat only well-cooked pork.

Typhoid Fever

Typhoid fever has largely disappeared in the United States as a waterborne disease and yet it has not been completely eliminated, even though the concentration of typhoid bacteria in existence has been tremendously reduced. Periodic outbreaks of typhoid fever have resulted largely from contamination of food by a typhoid carrier employed as a food handler. There are many people who carry typhoid bacteria but do not suffer from typhoid fever. It is important that the typhoid carriers do not have an opportunity to pass the typhoid bacteria on to other people. For this reason typhoid carriers are prohibited by law from being employed as food handlers where the opportunity for spreading the disease to unsuspecting persons would be rather high. Because of the importance of typhoid carriers as disseminators of typhoid fever, very close check is kept on the known typhoid carriers.

Shellfish have in the past been implicated in the spread of typhoid fever and other enteric diseases. Shellfish can become disease carriers since they are often consumed raw. The shellfish normally become infected as a result of being grown in sewage-contaminated waters. Regulations controlling the growth of shellfish in nonpolluted waters have greatly reduced the spread of disease by this source. In some areas the harvested shellfish are placed in large tanks containing chlorinated water. The chlorinated water passes through the shellfish and kills any pathogenic bacteria. The chlorine is purged from the shellfish by allowing it to live in clean water for a short period. The proper control of shellfish sanitation has all but eliminated shellfish as a vector for enteric diseases.

Salmonellosis

While typhoid fever is caused by *Salmonella* and hence must be considered as a form of salmonellosis, there are a large number of other *Salmonella* which can be transmitted through food and cause disease. Most *Salmonella* are spread by carriers, but the carriers are so numerous and difficult to detect that they cannot be controlled as easily as the typhoid carriers. The speed of salmonellosis can be reduced by proper sanitary precautions employed by the food handlers.

Amebiasis

Amebic dysentery is a common disease spread by uncooked vegetables grown in areas polluted with raw sewage. The causative agent for amebiasis is *Entamoeba histolytica*, a protozoa which forms quite resistant cysts which enable it to survive under adverse conditions. Amebiasis is rather rare in the United States as a result of sewage treatment and control of sewage effluents as irrigation waters. In some countries where the sanitation is poor, 50 per cent or more of the people are affected by amebiasis. The best control for amebiasis is cooking of the food and the prohibition of the use of sewage to fertilize food crops which are consumed raw. For the most part amebiasis is transmitted through salads. The American tourist is unusually susceptible to amebiasis as the result of his travels to countries which have lower sanitary standards and his insatiable desire for fresh salads. Amebiasis is one of the major causes for the lethargy of many of the inhabitants of Central and South America.

Food Poisoning

Some food poisoning is not caused by a direct infection of man by disease-producing microorganisms, but rather is caused by toxemias produced by microorganisms prior to ingestion. The most common toxemia is produced by *Staphylococci*. The *Staphylococci* will grow readily on any rich food such as the cream fillings of some desserts, egg salad, and custards. Proper refrigeration and rapid use of these foods can retard the growth of *Staphylococci*. As always, proper food preparation is essential to reduce the contamination to a minimum.

Botulism is produced by a neurotoxin from *Clostridium botulinum*. The *Clostridium* are anaerobic sporeformers which appear frequently in improperly canned food. The controlled methods for preparing commercial foods has all but eliminated botulism from this source but home-canned food has permitted botulism to remain as a danger. *Clostridium botulinum* appears to grow best in foods high in protein such as beans. Fortunately, the neurotoxin is heat-sensitive so that proper cooking of the foods after being removed from the can can destroy the toxin. Education of the home canner appears the best control for botulism.

RESTAURANT SANITATION

The large number of people who depend upon restaurants each day form a reservoir for the spread of the diseases just discussed. Therefore it is one of the duties for local health departments to inspect restaurants regularly and to ensure that all sanitary regulations are obeyed. A critical point in the potential transmission of microorganisms from one person to another person is with the utensils which must be used over and over again. It is necessary that all glasses, plates, and silverware be completely cleaned and disinfected. In large restaurants the cleaning procedures are largely mechanical with each step of the cleaning process carefully controlled. The only problem with this equipment is to make certain that it is operating according to specifications.

The majority of restaurants are small, and the dishes are washed by hand and returned for reuse as fast as possible. It is in the small operations where there is danger of contamination being transmitted and where supervision must be continuous. The proper cleaning process consists in scraping the excess food from the utensil into garbage cans; pre-rinsing to remove the small food particles; washing with a heavy-duty detergent; rinsing to remove the detergent and fine organic matter; and a final bath in a sanitizing solution. It is possible to combine the detergent and sanitizing step into a single operation. The quaternary ammonium type detergents are used quite extensively in dishwashing as a result of the bactericidal properties of quaternary ammonium compounds. In some instances the quaternary ammonium compound is used following washing with anionic detergents. Chlorine compounds are also used as sanitizers in many instances. After the final bath the dishes are allowed to drain dry and are not to be wiped with a towel. Often the final rinse is a very hot rinse so that the heat of the dishes will cause rapid evaporation of the moisture and permit rapid reuse of the dishes.

The efficiency of any cleaning process and of the sanitizing step can be determined only by bacteriological analysis. Normally, the bacteriological samples are collected at random intervals so that the food handler cannot take special precautions on the day the inspector arrives. Sterile cotton swabs are used to remove the bacteria from the surface of the eating utensil. On return to the laboratory the swab is placed in sterile dilution water and agitated so that the microorganisms will be transferred from the swab into the water. Serial dilutions are then made for total bacteria counts. Excessive bacteria counts are indicative of poor cleaning procedures.

The importance of proper cleaning techniques cannot be overemphasized in controlling the spread of infection. Improper cleaning can leave a thin organic film on the utensil which can serve as a breeding ground for bacteria. A good example of this was a case in the military service where vacuum bottles were used to transport coffee and cold water alternately. Although normal cleaning procedures were used on the vacuum bottles after each use, the coffee left a thin coating of oil on the inside of the vacuum bottle which was not removed by the cleaning procedure. Use of contaminated ice once permitted the enteric pathogens to gain a foothold in the thin oily layer and to grow. The cleaning procedures did not kill the enteric bacteria so that they were able to survive and grow in the oil film, continuing to infect the users of the vacuum bottles long after the initial contamination. Only after exhaustive search of all possible sources of contamination was the vacuum bottle uncovered. Normal washing procedures resulted in the vacuum bottles being rejected as a source of contamination until all other possible sources had been eliminated. Changing the type of detergent used to an ethanolamine type resulted in complete removal of the coffee film from the vacuum bottles and solved the problem of chronic intestinal disorder in the men at this post.

Quality Control

The role of the sanitary microbiologist in the sanitary quality control of milk and food is not a glamorous job but is an essential job for safeguarding the health of the people. It is primarily a routine analysis which must be made regularly to ensure that disease transmission through milk and food are kept to a minimum. The periodic outbreak of an unusual infection as in the case illustrated above can require extensive detective work on the part of the sanitary microbiologist before he finds the cause of the infection and eliminates it. With the increased population and the complexity of modern living the sanitary microbiologist is going to find an ever-increasing role in milk and food sanitation control.

SUGGESTED REFERENCES

1. "Standard Methods for the Examination of Dairy Products," 10th ed., American Public Health Association, New York, 1953.
2. Hilleboe, H. E., and G. W. Larimore, "Preventive Medicine," W. B. Saunders Company, Philadelphia, 1959.

CHAPTER 27

Air Microbiology

The air we breathe is far from being a sterile medium, but fortunately it is not a medium in which microorganisms can grow. If air were completely quiescent, we would see the suspended matter drop out of the air; but because of wind the movement of air permits large quantities of suspended solids to be retained for long periods of time. These suspended particles often contain microorganisms so that the wind can carry the microorganisms for long distances.

Types of Microorganisms

Bacteria, fungi, and viruses are the predominant microorganisms found in air. The lack of nutrients prevents survival of the vegetative cells for long periods but the formation of spores by certain bacteria and by fungi permits their survival for very long periods. Vegetative cells must have moisture in order to remain alive. Moisture in the air usually results from aerosols such as a person produces when he sneezes or even when talking. The aerosol usually evaporates rather quickly so that the vegetative cells do not spread far from their source.

The spores are protected against drying and can carry for long distances from their source. The dust storms in the western part of the country are responsible for the movement of large numbers of fungi. One of the problems a microbiologist faces in a dusty area is air contamination. Even the most careful microbiologist will have some extraneous contamination from the microorganisms contained in the air. Since viruses do not metabolize outside the host organism, it is not surprising to see that they are able to survive in air for long periods. The small size of the viruses permits them to remain in suspension for long periods of time.

Air-borne Diseases

The common diseases which have been found to be spread through the air include the following:

1. Diphtheria—*Corynebacterium diphtheriae*
2. Scarlet fever—*Streptococcus scarletinae*
3. Tuberculosis—*Mycobacterium tuberculosis*
4. Pneumonia—*Diplococcus pneumoniae*
5. Whooping cough—*Bordetella pertussis*
6. Smallpox—virus
7. Chicken pox—virus
8. Measles—virus
9. Mumps—virus
10. Influenza—virus
11. Common cold—virus
12. Poliomyelitis—virus
13. Systemic mycosis—(fungi) *Candida albicans* and *Histoplasma capsulatum*

It can be seen that many of the more common diseases facing man today are air-borne diseases.

Control of Air-borne Diseases

One of the most difficult problems in sanitation today is control of air-borne infections. There is no way in which air can be treated to destroy all the microorganisms which it contains. The tendency for people to work in close contact with each other in large cities makes air-borne infections very easily spread. This can be attested to in the recent epidemics of Asian flu which moved from city to city infecting great masses of people.

Sunlight is the most potent weapon against air-borne infection. Ultra-violet radiation from the sun destroys many of the microorganisms before they reach a potential host animal. Because of the germicidal properties of ultraviolet light, it is used in many areas requiring germ-free atmosphere. Unfortunately ultraviolet has poor penetrating powers so that it can be used only in small areas. The fact that ultraviolet radiations are harmful to people in excessive dosages means that proper precautions must be taken to prevent excessive exposure.

Chemical agents such as propylene glycol and triethylene glycol have been sprayed in the air to disinfect a contaminated area. Inside a building the air-borne contamination is largely related to dust and dirt stirred up by people walking. Disinfection of the floors at periodic intervals can effectively reduce the bacterial population in the air. Regular cleaning of floors is important in keeping the air-borne contamination to a minimum.

Filters can remove microorganisms as well as dirt from the air. Cotton plugs on tubes of bacteriological media testify to the effectiveness of cotton as a biological filtering material. Glass fibers are also used as filters. The problem with filters is the rate at which they clog with solid particles. Recently electronic precipitators have been used as air filters. The air is passed by charged plates having a potential of several thousand volts. The charged particles in the air are attracted to the charged plates and thus are removed from the air. When the plates become clogged with particles, the electrical potential is removed and the plates are washed with hot water. After the plates have dried, the electrical potential is reapplied and the electronic precipitator is ready for reuse.

Water has been used as an effective filtering agent for air-borne particles. The air is passed through a moving curtain of water which traps the particles and thus removes them from the air. Some air-conditioning units use water sprays to reduce the air-borne contamination.

Sampling Air

The best qualitative test for air-borne contamination is the exposure of a petri dish containing a suitable medium to the air for a few minutes. The microorganisms in the air will settle onto the nutrient medium and will grow surface colonies in a short incubation period. The growths on the plates demonstrate the presence of fungi, actinomycetes, yeast, and bacteria.

FIG. 27-1. Wells air centrifuge for determining the bacteria in air.

Quantitative measurements of air microorganisms require the passage of a known volume of air through an entrapping fluid over the surface of the nutrient medium or through a bacterial filter. The liquid-entrapment method depends upon bubbling the air through the liquid, usually dilution water, and subsequent serial dilution transfers to nutrient media for culturing and counting. The problem with the liquid-entrapment method is the failure to remove all the microorganisms by the liquid. Usually the air is introduced at the bottom of a packed column of sterile glass beads so that the turbulence of passing through the packed column will produce sufficient contact between the water and the microorganisms to ensure their removal.

The impingement method requires directing the air against the nutrient medium with sufficient force to cause the suspended particles to impinge upon the nutrient medium as the air passes over its surface. The Wells air centrifuge shown in Fig. 27-1 is an impingement device which utilizes centrifugal force to impinge the microorganisms against the nutrient medium which lines the walls of a special glass counting tube. The Wells air centrifuge is very efficient in removing the microorganisms from the air. Figure 27-2 shows the microorganisms grown on agar-coated tubes from the Wells air centrifuge.

One of the newest methods for removing microorganisms from air for quantitative enumeration is the membrane filter. The membrane filter retains all the microorganisms in a given volume of air which is passed

Fig. 27-2. Microbial growths on agar tubes used in Wells air centrifuge at two different air volumes.

through the filter. The membrane is removed from the sampler and transferred to a nutrient medium and incubated. The simplicity of the membrane-filter technique makes it quite useful in air microbiology. Since the same device can be used in studying particulate matter in the air, the sanitary engineer should become quite familiar with use of the membrane-filter apparatus.

Microbial Content of Air

Air microbiology has received relatively little study as compared with other phases of microbiology. The possibilty of biological warfare (BW) using air-borne infections has given considerable impetus to air microbiology with much work being done at Camp Detrick. The variability of the biological content of air with atmospheric conditions makes quantitative enumeration of bacteria of limited use.

A study of the microbial content of air in a hospital at Lynn, Mass., has shown a bacterial content of from 3 to 20 bacteria per cubic foot of air. Identical analyses made at the Lynn City Hospital 4½ years apart are given in Table 27-1. The initial studies were made to show the effec-

TABLE 27-1. MICROBIAL COUNTS IN AIR AT LYNN, MASS., HOSPITAL

| Location | Microorganisms in air, counts/cu ft | | | |
| | 24 hr | | 48 hr | |
	1955	1960	1955	1960
Operating room:				
Fresh air	1.3	1.1	1.8	1.3
Return air	3.4	2.0	3.6	2.4
Operating room	3.2	1.7	3.2	2.1
Third floor:				
Fresh air	0.7	1.6	1.3	1.9
Return air	0.7	1.9	0.8	2.3
Room (empty)	1.8	2.8	3.1	3.2
Children's ward:				
Fresh air	2.7	1.5	3.0	1.7
Return air	0.8	1.4	2.0	1.6
Room (occupied)	2.0	2.9	3.2	2.9
Maternity delivery room	6.4	0.6	7.3	0.7

tiveness of electronic precipitators in removing bacteria from air recycled from the hospital back through the air-conditioning system. These studies were made in a new wing of the hospital just after it was completed and placed into operation. The second studies show the effectiveness of the system in maintaining high-quality air even though potential air contamination might be considered as high. It demonstrates that air from a hospital can be recycled in part to achieve economy in operation of air-conditioning equipment without creating a health hazard.

Effect of Precipitation

Precipitation in the form of rain, snow, sleet, or hail is very important in cleaning the air of contamination. The initial period of precipitation results in high bacterial contamination, but the later precipitation is relatively free of contamination, testifying to the effectiveness of precipitation as an air-cleaning device. Where water is collected from the roofs of buildings for storage in a cistern and human consumption, the system should be so constructed that the first part of the rainfall is diverted from the cistern to allow the contaminated water to be discharged away from the cistern. In this way the water in the cistern can be kept of relatively high quality.

As modern technology increases control of the atmosphere in which we live, there will be a greater effort to understand the various aspects of air microbiology so that transmission of air-borne infections can be reduced to a minimum. Air conditioning of individual homes and even of entire cities will permit the sanitary engineer to exercise control on the spread of microbial infection through the air in the same way that he has controlled the spread of infection through water and food.

Index